Ecological Principles
for
Economic Development

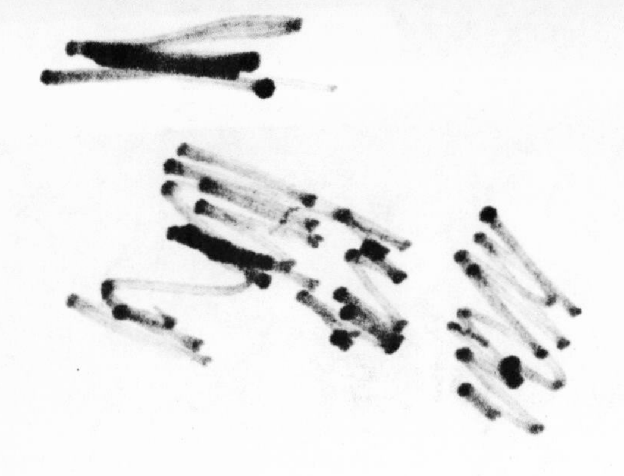

Ecological Principles for Economic Development

RAYMOND F. DASMANN
JOHN P. MILTON
PETER H. FREEMAN

Published for
The International Union for Conservation of Nature and Natural Resources (IUCN), Morges, Switzerland
and the
Conservation Foundation, Washington D.C., USA

JOHN WILEY & SONS LTD
London · New York · Sydney · Toronto

Library of Congress catalog card number
72-8597

ISBN 0 471 19606 1

Reprinted July 1974

Printed in Great Britain
by Unwin Brothers Limited
The Gresham Press, Old Woking, Surrey
A member of the Staples Printing Group

Foreword

Background

This book had its origin with several years of research culminating in a conference sponsored by the Conservation Foundation of Washington D.C. and cosponsored by Washington University of St. Louis, which was held at Airlie House, Virginia, in December 1968. This conference, on *The Ecological Aspects of International Development*, brought together representatives of international development agencies and of geographical areas undergoing rapid development, with experts having wide experience of ecological problems associated with development activities. It featured the presentation of case histories of the environmental and social consequences of past efforts at economic development. The discussions brought out the necessity for closer relationships between development agencies and those trained in the techniques of ecological evaluation and environmental assessment. The case histories and conference discussion have been published as *The Careless Technology: Ecology and International Development* edited by M. T. Farvar and John P. Milton, Natural History Press, Doubleday & Co., New York, 1972.

Following the conference there were many meetings between those associated directly with economic development and those concerned with ecology and conservation—including representatives of the World Bank, UNESCO, FAO, the UN Development Programme (UNDP), IUCN, the International Biological Programme (IBP) and the Conservation Foundation. On the initiative of Mr. E. M. Nicholson of IBP, it was decided in 1970 to reassemble a small group of experts from development agencies, various countries and the environmental sciences, to consider how further progress might usefully be made.

At FAO headquarters in Rome during September 1970, representatives of the World Bank, FAO, UNDP and United States and Canadian international development agencies met together with those of the Conservation Foundation, IUCN, IBP and other ecologically-oriented groups. Representatives were also present from India, Thailand and the UAR. One of the conclusions reached was that the moment was opportune for IUCN and the Conservation Foundation to cooperate in the preparation of an *Ecological Guidebook for Economic Development*. Its aim would be to provide ecological guidelines for use by development planners and decision makers and also to explore the pertinent interrelationships between economic development, conservation and ecology. The meeting felt that such a book could be of great value to those concerned with development, particularly at the international level, in reaching decisions on future projects.

Accordingly, Dr. R. F. Dasmann of IUCN and Mr. John P. Milton of the Conservation Foundation were asked to prepare a first draft for consideration and comment by development agencies and other expert organizations and individuals. Mr. Peter Freeman was subsequently commissioned by the Conservation Foundation as an additional author. After further interagency consultation, it was agreed that the initial draft should be limited in scope (see below) and place emphasis on ecological principles and theories of vital relevance to international development. It was also considered that, taking account of the forthcoming publication of the Conservation Foundation's book *The Careless Technology*, which included the case histories presented at the Airlie House Conference, further research into the history of past development projects should not be undertaken for the purpose of the present enterprise but deferred for a later occasion.

Scope

The book is written from the point of view of the ecologist for the use of those concerned with development whether at a purely national level or in connection with any of the aid programmes of the international agencies. It sets out to explore as briefly as possible some of the ecological concepts which have been sufficiently tested in practice to be considered valid and useful in the context of development activities. Particular emphasis is placed on ecosystems which are currently subject to heavy development pressure: for example, those of tropical humid forests and of savannas and grasslands in the tropical, sub-tropical and mediterranean regions; and those known to be especially fragile, such as high mountains, coastal areas and islands. By way of leading into future studies, various problems associated with dams, irrigation and other major river basin development, forestry, livestock and agricultural projects, and the rather specialized case of tourism development, also receive attention.

No attempt is made at comprehensive treatment of the ecology of any of the ecosystems concerned; to do so would have required several more volumes, even if in some cases relevant studies have already been published. Similarly, no complete case histories are included, but several are summarized

and reference is made in the bibliographies to sources of relevant case material. In short, this is an exploratory volume and not intended to be a final product. If it proves useful to those who deal with development planning and execution, it can in due course be built up and developed, on the basis of new ideas and suggestions, into an increasingly comprehensive and valuable reference book.

Limitations of Knowledge

It is customary in books concerned with ecology to point out the need for further research. This need is apparent, but should not be overemphasized. The greater immediate need is for effective application of ecological principles that are already known. Enough ecological knowledge is now available to permit a far better job to be made of development than ever in the past.

It became clear during the preparation of this book that much can be gained through an environmental evaluation of past development projects, aimed at discovering the reasons why some have succeeded while others have sometimes failed. It is perhaps of special importance to evaluate systems of stable and profitable use of land and resources, particularly those that have continued over centuries without serious impairment of productivity or loss of environmental values. Despite the expenditure of huge sums of money, internationally, on a wide range of development projects, there is still too much reluctance to look back, evaluate past efforts and draw up detailed balance-sheets of costs and benefits, both economic and environmental, particularly on a long-term basis. To implement such evaluations means bringing together the facts we have learned and discovering the reasons why things so often turn out differently from what was intended. This comprehensive post-audit analysis is largely beyond the scope of the present volume but should be a goal for future work within development agencies and developing regions.

No claim is put forward here that ecology is a touchstone for success in international development, where economics and engineering have often failed. But just as it has long been obvious that development efforts which ignore economics and engineering are likely to founder, so it should by now be equally obvious that development efforts that take no account of the ecological 'rules of the game' are also bound to suffer adverse consequences. Although not stressed in this book, examples are also quoted to show that ignorance of human behaviour has been a further reason for lack of success. Hopefully, through this and future work, a truly interdisciplinary approach to international development can be achieved that will permit mankind to reach the high quality of life to which all people aspire.

Acknowledgements

Preparation of the first draft of this book was carried out by R. F. Dasmann, Senior Ecologist of IUCN, John P. Milton, Acting Director of International Programmes, and Peter Freeman, Consultant, of the Conservation Foundation, with the active collaboration of Gerardo Budowski, IUCN, and Thane Riney, FAO. Research assistance was provided by K. MacNamara of the Conservation Foundation. Various ideas and assistance were provided by E. M. Nicholson and E. B. Worthington, International Biological Programme; M. Batisse, UNESCO; Frank G. Nicholls, IUCN; and Gordon Conway, Imperial College of Science and Technology.

The initial draft was reviewed by approximately forty persons with knowledge and experience in economic development projects. Their advice was most valuable for redrafting purposes and in the preparation of the final text by Sir Hugh Elliott, IUCN's Scientific Editor. IUCN and the Conservation Foundation express their sincere gratitude for the assistance provided by the named individuals and by the many and necessarily anonymous reviewers, as well as by all the organizations which have so readily collaborated in bringing the work to its conclusion.

Contents

Summaries of the Chapters

Chapter 1: Introduction—Goals of Development and their Attainment

Conservation and economic development should ideally be directed towards a common goal—the rational use of the earth's resources to achieve the highest quality of living for mankind. In practice, economic development tends to place stronger emphasis on quantitative increases of production, aimed at enhancing the material well-being of people, whereas conservation, while concerned with sustaining quantitative yield, also emphasizes management of more qualitative aspects of the human environment which can add depth and meaning to human life.

With both conservation and economic development it is essential to consider the physical and biological rules within which all life on earth must operate. These are the subject matter of the science of ecology which is the study of the relationships of organisms with their environment. Proper consideration of ecological principles will assist those concerned with development or conservation to achieve their goals with a minimum of undesirable side-effects, reducing the likelihood of major environmental disturbances which could be harmful to all life within a region or throughout the world. Lack of consideration for the ecological realities of an environment can doom development efforts, with consequent waste of money and impairment of the conditions of life, just as surely as if the technological, economic, political or social factors were to be ignored.

Conflict between conservation and development can be minimized if there is an understanding and partnership among those whose main interests lie on either side, particularly during the process of selection from the half dozen basic options that are open when land use of new development areas is under consideration. Developers, on the one hand, must have due regard for environmental values, the conservation of which is important for scientific, recreational or other reasons not easily capable of measurement in precise monetary terms. This may include the reservation of representative and sometimes extensive wild areas, with their full array of plant and animal species, for purposes ranging from detailed scientific research to leisure-time enjoyment; purposes which can in themselves become major economic assets.

On the other hand, those concerned with conservation must be equally ready to recognize the political, social and economic forces behind the development drive and be prepared to give every assistance in exploring alternatives and reaching reasonable compromises, whenever interests are in conflict. In resolving the latter, three ecological principles are of special relevance: first, the need to keep a range of resource use options available to future generations; secondly, the fact that, for reasons based on their history and evolution, more intensive agricultural and pastoral development of lands of proved productivity is likely to give a better return than attempts to develop marginal areas; and, thirdly, the fact that conservation of species and natural communities is a logical first step in development, for the reason that the resources of such communities are irreplaceable both from the point of view of satisfying human needs and aspirations and also because of their long-term contribution to the stability and productivity of the planet.

Chapter 2: General Ecological Considerations

All economic development takes place within natural ecosystems, which may or may not have been already modified by man. Development brings about very varying degrees of modification, but always remains subject to the ecological limitations which operate within natural systems; these limiting factors must be taken into account if the development is to succeed. The total complex of ecosystems is known as the biosphere, because it comprises the surface layer of the planet capable of supporting life. If the rules which govern the health and functioning of the biosphere are ignored life must cease. The ecological principles related to this functioning apply equally to all development situations.

The plants and animals able to exist in any area will form a biotic community and an ecosystem is, essentially, a biotic community in interaction with its physical environment of sunlight, atmosphere, water and soil or rock. Within the ecosystem each species exists as a population, the growth or decline of which is affected by the capacity of the system to provide the requirements of life. It follows that although man may modify or even destroy particular ecosystems, he cannot bring about major modifications of the biosphere as a whole without risk to his own survival. It also follows that there must be limits to population growth of all plants and animals including man. As populations approach the level at which the environment is capable of providing no more than bare subsistence, conditions of life worsen for each individual. Beyond this point, population growth must come to a stop either through curbing of birth-rate or an increase in deaths, which in the case of man implies the exercise of choice and of voluntary limitation.

The environmental limits to growth determine the *carrying capacity* for any species, which may be at a subsistence level, a security level (implying reasonable freedom from privation, disease or predation) or the optimum level, which is the normal objective for human populations, their domestic animals and their crops. These levels are governed by a variety of *limiting factors*—climate, soils, water and the complex of biotic factors which includes the transmission or cycling of energy and nutrients, and the impact of parasites, diseases and other kinds of predation.

Of particular importance is the manner in which these factors interact. Such interaction is a frequent cause of failure in economic development, when efforts are made to modify one limiting factor (e.g. lack of water in an arid region) without considering its relation to other factors (e.g. the food supply in the area where additional water is being provided). Because of its relation to efforts to enhance agricultural productivity, another important influence on the situation is the way in which *food chains* operate. The biological processes involved mean that substances such as radioisotopes or pesticides, even if present in only small amounts in the general environment, are concentrated as they pass from soil to plant to herbivore to carnivore and may reach unacceptably high levels in terminal species, which notably include man.

The value of maintaining a diversity of species which can enhance the stability of a natural or man-modified ecosystem (e.g. by preventing outbreaks of pest species which may otherwise have to be controlled by chemicals, leading in turn to new problems as resistance is built up), is of particular significance for development. If ignored, consequences may ensue which are the reverse of what was intended when a project was planned. A favourable factor in this situation is the process of biological succession and the resilience of natural systems. But due to isolation or rigorous climate, and consequential lack of diversity, some ecosystems, notably arid or cold regions, oceanic islands and high mountains, are 'fragile' and great care is needed in effecting any modification. In fact any ecosystem can be pushed to a 'point of no return', a threshold beyond which the limiting factors are so severe that biological repair becomes, in human terms, an intolerably slow process.

Chapter 3: Development of Humid Tropical Lands

The tropics, particularly the forested humid tropics, have been resistant to efforts at economic development. With some notable exceptions, such as the alluvial and younger volcanic soil areas of South-east Asia and of a few localities in Africa and Latin America, most tropical regions support small human populations and have contributed little to human food supplies. The ecological reasons for this stem largely from peculiarities of climate, soil and biota. They are reflected in the success or otherwise of ways of utilizing tropical forest regions which have been practised or applied in the past, including flood-plain cultivation, terracing, shifting cultivation and various methods of refining or replacing the original forest cover.

A particular feature of the humid tropics is the enormous diversity of life. This is both a benefit to man if properly used and also an obstacle to economic development. The complexity of interactions between soil, climate and the great numbers of plant and animal species, contributes to the stability of the forest ecosystem under natural conditions. But it militates against successful agricultural or pastoral development, or rapid recovery if disturbance has been too prolonged or acute. The frequently experienced inability of certain tropical soils to sustain intensive agriculture points to the need for concentrating development in areas where soils are more suitable, but which are at present poorly managed and producing below capacity.

The special values of natural primary forest, and in particular of highly diversified rainforest, for environmental, scientific, educational and recreational purposes suggest that, in situations often met with in the humid tropics, a reversal of customary development priorities will often be appropriate. Thus, the first concern would be to protect extensive and well-distributed areas in their primitive state, in which they can nevertheless contribute in many ways to the economic well-being of the regions concerned. This by no means precludes, within the forestry context, the expenditure of effort on establishing plantations and intensive management of secondary forest, both of which can greatly increase the flow of forest products. Similarly, despite the ecological limitations, the proper development of agriculture, especially of tree crops, can increase the capacity of these regions to support human populations at higher material levels. Emphasis needs to be placed on the better soils and on the gradual replacement of the sheltering and soil-renewing properties of natural forest through the adoption of agricultural practices, including balanced use of fertilizers, which provide equivalent protection. This does imply, however, a fairly high level of education and technical expertise among the farmers concerned.

Unless such developments are methodically planned, however, there is a constant danger of accelerated erosion, laterization of certain soils or the rapid loss of their structure and fertility, and, in the case of extensive monocultures, persistent difficulty with animal, plant and microbial pests, often merely complicated by recourse to chemical controls. This is apparent in over-hasty attempts at pastoral development, involving the substitution of grassland for humid forest, which is a widely practised but usually short-lived activity, leading to the creation of unproductive scrub and abandonment.

Other problems, which are not peculiar to development in the humid tropics but tend to have specially undesirable consequences, are those associated with the use of pesticides and herbicides and with the opening of access to new regions by road construction. Unless the consequential movements and land-use practices of people can be effectively controlled, both these forms of development may prove to be counter-productive, leading to random destruction of lands and resources that it would have been better to have reserved for more valuable long-term use.

Chapter 4: Development of Pastoral Lands in Semi-arid and Sub-humid Regions

The development of rangelands offers a great opportunity for enhancing the well-being of peoples in many parts of the world, notably in the sub-humid and semi-arid tropics and sub-tropics. Yet there are few fields of economic development in which greater or more destructive blunders have been made than those involving attempts to improve the productivity of pastures. Although educational, social and political factors are often implicated, an underlying cause has been the lack of understanding of the principles of rangeland ecology.

Among the ecological characteristics of drier rangelands, climatic variability and its effect on soil and vegetation are of prime importance. It means, for example, that management must often be based on expectation of drought. This in turn implies special attention to conservation of soil and grass cover and, in particular, to the control of numbers and distribution of grazing animals.

Fire is both a natural recurrence and a useful management tool in dry rangeland. It can be employed to eliminate undesirable vegetation, improve pasture quality and influence the distribution of animals. The dangers inherent in its use are least evident in sub-humid savanna, but everywhere demand the most careful study of frequency, seasonality and intensity of burning, if effects opposite to those intended are to be avoided.

Although, in the production of protein of value to man, grasslands and savannas have an outstanding potential, their productivity varies considerably, declining generally from forest edge to the arid extreme. The economics of intensive management are therefore equally variable, so that, for example, in many areas utilization of wildlife resources may be capable of producing more income than could normally be obtained by concentrating entirely on domesticated species. Thus, desertification has often accompanied the use of more arid rangeland for livestock. Control of their movements and numbers may be essential to halting this process and restoring productivity.

The condition of most rangelands, under current patterns of use, tends to be unsatisfactory, but is seldom beyond repair. The prerequisite of remedial development, however, is the ability to assess condition and trend, to recognize and reverse deterioration before it becomes too severe. Techniques for doing so are available and are based on the evaluation of evidence from changes in soil structure and in the distribution, species composition and physical state of vegetation and of wild and domestic herbivores. Although these techniques must be expertly designed and adjusted to local circumstances, they can easily be learned and applied.

The greatest single factor contributing to rangeland deterioration tends to be the mismanagement of livestock. Unless there is a real possibility of rectifying this, efforts and expenditure on range improvement will be wasted. Assuming, however, that effective grazing control can be ensured, major gains in productivity may be made by means of techniques such as fertilizing, reseeding, water development, disease control and, as previously mentioned, the use of fire. In each case, precautions against bad planning and misapplication are vital for success. Once again, it is emphasized that greater returns will usually be obtained by concentrating on better sites, and upgrading them from a depleted to a highly productive condition, than from equivalent expenditure on establishing livestock in more marginal rangelands better suited to other forms of development.

Chapter 5: Development of Tourism

One of the possible approaches to the economic development of any region, referred to briefly at the end of the previous chapter and involving some rather special ecological problems, concerns the utilization of tourist potential. Partly, no doubt, because it is not directly related to commodity production, it is only in quite recent years that this has attracted the attention it deserves. Yet in the form of 'international tourism', aimed at holiday-makers, enthusiasts and students of every country, it is now widely recognized to be a major source of employment and income.

Two considerations are peculiar to tourism development and need to be taken into account when integrating it, as it should be integrated, with any overall regional development plan. The first is that, despite the demand for various sophisticated facilities which often accompanies it, tourism is basically dependent on unspoilt environment. In most other types of development, some environmental values have to be sacrificed in return for expected benefits, but for tourism the maintenance of these values at a high level is essential. Well-planned tourism can in fact help both to justify and safeguard the quality of the environment.

Secondly, however, the infrastructure of tourism—hotels and lodgings, roads, vehicles, power lines and the rest (not excluding the problems of waste disposal generated by them)—may, like that of any other industry, bring about deterioration; but it does so to an unusual extent in ways which affect its own survival. Varying with the fragility of the area concerned and the particular nature of tourist activities contemplated, there is a definite limit to carrying capacity for tourism. The first necessity, therefore, in tourism development is an evaluation of the resources and landscape available, in order that a reliable estimate of this capacity may be made.

The importance of these principles can best be appreciated in relation to problems of three of the major growth areas of current tourist activity—

national parks, coastal regions and islands (particularly the smaller oceanic islands of the tropics).

National parks are by definition intended to permit and encourage the continuing enjoyment and appreciation of nature and landscape. Although most kinds of development are, therefore, normally excluded or severely restricted, that of tourism is logically inherent in a parks system. Nevertheless, in a national park, as now generally defined, some modes of tourist activity are appropriate, others, such as those involved in mass recreation, much less so or not at all. It follows that the application of zoning principles to planning and management is of high importance. It can, for example, enable one park or sector of a park to be visited by very great numbers of people, while another serves a vital scientific function in protecting genetic stocks or 'gene pools' of plants and animals.

Apart from general management problems, there are several of a more specialized nature, such as those concerned with the difficult ethical, social and administrative questions raised by the presence of primitive peoples in some park or potential park areas, including islands. There are also special problems involved in tourist impact on more fragile ecosystems, such as those of high mountains, and on marine parks, a comparatively new development of which the full ecological implications have still to be learned. However, by far the most vulnerable of high-value natural areas at the present time are coastal regions (including estuarine mangrove as well as attractive beach and lagoon) and the terrestrial ecosystems of islands. Disturbances, originating from tourism development, which may affect these, include all those arising from construction, dredging, pollution, erosion and excessive population pressure. In the case of islands, the position is often aggravated by unwise or accidental introductions of species. There is a strong case for setting aside for science, under international convention, an adequate sample of the rather few islands or parts of islands which remain uninhabited and comparatively undisturbed. Directly or indirectly, this may also eventually serve the interests of tourism development.

Chapter 6: Agricultural Development Projects

The ecological significance of agricultural development is related to its main objective, increased production of food and fibre, and to the changes brought about in the agroecosystem by the techniques used to achieve that objective. An agroecosystem is the complex of organic and inorganic components of a region, as modified primarily by cultivation but also by all other human activities.

Agroecosystems may have a comparatively stable vegetational cover, for example coffee, or a discontinuous one such as an annual crop. In either case weed and insect pests are characteristic and liable to become critical or dominant features of the system, their explosive outbreaks indicating lack of stability. The latter also varies with plant diversity, the life-cycle of main crops, climatic fluctuation and the degree of isolation from other ecosystems. The trend, under development, is from diversity to simplicity, in which natural checks and balances are greatly reduced and continuous effort and expenditure are needed to replace them. The fully developed agroecosystem functions on the basis of imported energy, water, nutrients, chemicals, technical expertise, in exchange for which a major part of human needs, including even food, may be purchased out of income from production surpluses.

Traditional land use and farming techniques, such as shifting cultivation may have little effect on ecological stability, so long as populations of man and his domestic animals do not exceed the carrying capacity of the environment. But when they do and there has been consequent pressure to farm less suitable land, depletion of resources, malnutrition, erosion and eventual destruction, particularly of the marginal areas, have commonly occurred.

The introduction of modern production factors—mechanization, irrigation, fertilizers, pesticides and herbicides, and disease-resistant high-yielding crop varieties, not only radically alters ecosystem function and stability, but also frees the subsistence farmer from several limitations. However, instead of dependence on the food-producing capability of his land, he now becomes dependent on market and food distribution systems. Because of this and his weak competitive status, the farmer who is still essentially operating at a subsistence level may well need special support if he is not to overexploit his land, expand his activities into still more vulnerable areas, revert to primitive land use practices, or abandon his efforts and migrate to the town. All these choices involve deterioration of resources and there is a parallel risk of this in all the technical innovations which initiated the process, unless they are carefully controlled. Pesticides and herbicides are a notorious example and there is a strong case in favour of substituting for their unrestrained use integrated pest and disease control based on a combination of biological and chemical methods and soil and crop management techniques, and coordinated at an agroecosystem level.

This implies that the key factor must be agroecosystem management. In addition to integrated pest and disease control (with minimum use of ecologically disruptive persistent chemicals and careful attention to safeguarding and, if necessary, salvaging genetic materials), the goals of management include sustained productivity and the maintenance of maximum species diversity relative to natural diversity, of human and livestock populations within the limits of carrying capacity, of nutritional standards, and of soil and water conservation. Such management must depend on collection, analysis and interpretation of data, which should not as in the past place a too narrowly conceived emphasis on resource exploitation and investment justification. On the contrary, it is essential that the most broadly based ecological information should be made available to development authorities and project managers and kept continually up to date as development proceeds.

Chapter 7: River Basin Development Projects

Ever since agriculture began, men have tried to control the supply, distribution and quality of water. Ten per cent of the world's total stream flow is now regulated and regulation is rapidly being extended, particularly to less developed tropical regions and particularly in the form of reservoir and irrigation projects. In planning these projects too little account was taken in the past of social and ecological consequences and the main purpose of the chapter is to review potential ecological impacts of reservoir construction and irrigation, in the tropical context, and suggest principles and methods of improving future project planning.

Modification of a river ecosystem by construction of a man-made lake, in addition to inundating more or less extensive areas of land above the dam and affecting water flow, causes numerous other changes in the environment both up and downstream. Since many human communities and cultures have evolved social systems which are dependent upon the productivity of natural river ecosystems, they will be profoundly affected by the changes brought about by man-made lake construction, which may be either beneficial or disadvantageous. Provision of hydro-power, irrigation water, flood control, aquaculture and lake fisheries, or recreation facilities are clearly among the benefits. On the other hand, associated with each of these benefits a whole series of problems have to be faced and resolved, including human resettlement, pollution by pesticides and herbicides, aggravation of water-borne diseases, the spread of aquatic weeds and disruption of former riverine fisheries, one factor in which is the deposit of nutrients by sedimentation in the reservoir and consequent elimination or reduction in the supply downstream, with effects also on agriculture.

It is suggested in this chapter that through 'preventive planning', based on research, monitoring and analysis of the data supplied by many disciplines, the majority of adverse ecological consequences can be avoided, reduced or remedied in future tropical river basin development. This will often involve modification of project design to meet ecological needs, improved management to prevent disruptive ecological impacts and, therefore, the inclusion of alternatives in the cost-benefit analyses, so that choices can be made. Development projects which are otherwise sound ecologically as well as economically, may not be in conformity with the existing social system and culture. For this reason, planning should include strong emphasis on studies of the social background and on provision of extension services, so that to the extent that alternatives or adjustments are not available, or resettlement is involved, the effects of development will be beneficial in the long term.

Irrigation projects, which are often dependent on reservoir construction have one or two additional rather specialized problems. The most important of these is the potential for salinization, which is difficult and expensive to remedy, so that prevention based on adequate pre-investment surveys should always be the aim. Contamination problems tend to be aggravated by seepage into groundwater sources used for irrigation and also by direct application of chemicals to irrigated crops, but are otherwise similar to those which have to be resolved in other sectors of river basin development.

CHAPTER I

Introduction—Goals of Development and their Attainment

(1) THE NATURE AND AIMS OF ECONOMIC DEVELOPMENT

Ideally, economic development is a process through which nations seek to improve the well-being of their citizens. In any final summing-up the success of development should be judged on the basis of such improvement. In the following pages it will be assumed that this ideal goal of development is the true goal. It will be assumed that those charged with national leadership have the welfare of their people first and foremost in mind and that they have the power to control those whose means and ends may differ.

Much economic development may take place without having an obvious, direct effect upon the environment or its natural resources. For example, the establishment of an effective banking and credit structure in a country, the training and organization of administrators and managers, the building of a system of communication, and the establishment, staffing and organization of educational institutions, are as much a part of economic development as any manipulation of land and water. It is when economic development becomes employed in the exploitation of the environment and its resources that it becomes of direct concern in this volume.

At this stage in human history, the drive towards economic development is not to be turned aside. Indeed it is often given the force of a moral imperative for nations. Under the heading 'The duty to develop', the Advisory Committee on the Application of Science and Technology to Development of the United Nations Department of Economic and Social Affairs (1970) makes this statement:

'The resources of the entire world must be developed rationally—to the fullest extent possible with the means available; mankind as a whole can progress only by efficiently utilizing all of the earth's available natural resources, especially at a time when its population is growing at such a startling pace. No country can claim, either on moral or practical grounds, to be entirely independent from other countries, and the economists and scientists know that the fate of each is important to all—though not all people realize this. Each country bears the primarily moral responsibility for the conservation and rational development of its own natural resources. It is also the duty of every nation to participate according to its means in the development of every other nation according to the latter's needs.'

If in this book from time to time we question the development imperative, it is not because of opposition to development as a process but through concern with forms of development which have so often failed to produce improvements in the well-being of the people affected. It is to be hoped that countries and areas which still have a wide range of development options available will avoid the mistakes which have beset the progress of technologically advanced areas. Consideration of all the different values associated with the natural resources of any region may give pause to those who confine their attention to 'efficient utilization of all of the earth's available natural resources'.

Development is a dynamic process in which all the nations of the world are involved, and one in which there is as yet no recognized end point—no final level of achievement. The more technologically advanced nations continue to seek new means for improving their utilization of natural resources or for enhancing the conditions of the environment in which their people live. Less advanced nations strive to reach levels of economic well-being which the more advanced countries have achieved. Thus the categories of 'developed' and 'developing' nations have already been largely abandoned since all are developing but at different rates and from differing past levels of achievement. There are rich and poor countries, more or less advanced countries, according to various measurements, but the process of economic development affects them all. Furthermore it is no longer realistic to view the economic and ecological state of any one nation in isolation from all others. All are now part of an international community. All occupy the same biosphere. What happens to one affects the whole.

(2) THE RELATIONSHIP OF CONSERVATION AND DEVELOPMENT

With the appearance of a strong public interest in restoring the quality of the human environment and the planning of major national and international programmes directed towards this end, fears have been expressed of potential conflict between the interests of those concerned primarily with the conservation of the environment and those concerned with economic development. Although this apprehension has been strengthened by the results of some past conflicts between conservationists and development economists and by irresponsible statements from both sides, there is little reason for its continued existence. Properly interpreted, the goals of conservation and those of development should coincide if the long-term well-being of the human race is given equal consideration with the immediate needs of today's population.

The United Nations' definition of conservation, put forward by UNESCO and FAO, is—'the rational use of the earth's resources to achieve the highest quality of living for mankind.' This would be an equally good definition of the ideal goal of economic development. Thus there should be a growing convergence rather than conflict between conservation and development aims. However, it must be recognized that the concept of what constitutes 'quality of living' will vary with the cultures and the economic status of the people concerned. The term 'rational use' also varies in interpretation, since it may seem entirely rational to one group of people to mine an area even if this means the sacrifice of all other values, whereas to others rational use would call for the exclusion of mining in order to protect other resources or values. Nevertheless, such disagreements are likely to occur *within* the ranks of those labelled as 'conservationists' or 'developers' as well as *between* these two groups.

In practice, those whose prime consideration is economic development of resources will often place a strong emphasis on quantitative production—number of new acres brought under irrigation, increased yield of wheat or rice per acre, tons of minerals marketed, numbers of visitors to national parks, etc. These are direct and readily measurable economic gains. Conversely those more concerned with the conservation of the environment will look at direct and indirect socioeconomic costs, in the short and long term, as well as at the immediate benefits. This may mean less emphasis on quantitative production, while recognizing its importance, and more inclination to ask what desirable natural processes may be disrupted or what resources and values sacrificed in the effort to attain this increased quantity. Another basic question will be whether or not the people concerned might not benefit more in the long run if those processes, resources or values were maintained or at least modified in such a way that their integrity is maintained in part. Such differences in emphasis might lead to conflict only if one side is unwilling to examine alternative approaches to achieving the common goal, or if one side is unwilling to accept reasonable limits on its activities. Indeed, the modern economist is generally aware of social and environmental constraints upon development. Only recently, however, has he realized that a proper assessment of these constraints brings into play complex, technical disciplines with which he is usually poorly equipped to cope. Economics has outrun the other disciplines in developing an intensive, specialized approach to international development. But it is now essential (and overdue) that other ways of looking at the problems be equally stressed.

It is naïve to believe that disagreements can always be minimized or readily resolved. It is reasonable, however, to believe that a joint understanding of the 'rules of the game', as they appear to the participants in any disagreement, will relieve much controversy.

(3) FACTORS TO BE CONSIDERED IN DEVELOPMENT PLANNING

The decision to develop a particular area or natural resource is legally an internal decision within a nation made by a government, or by private groups with the consent of government. The decision may be strongly influenced by the advice of agencies external to the country concerned, and the development may be carried out by a foreign group, but it remains a national prerogative. As such, it is commonly motivated by political considerations as well as by economic factors. Generally speaking, there are many alternative ways in which money can be invested in the expectation of enhancing the economy of a nation. The decision to pursue one way in any area is most commonly politically motivated. For example, many of the ecologically and socially unwise decisions for the development of semi-arid rangelands have resulted from conditions of political unrest involving nomadic or semi-nomadic pastoralists. In the absence of this political pressure it is doubtful if economic justifications or engineering feasibility studies would have carried much weight.

The political drive behind development is often particularly difficult to deflect by scientific analyses or the statistics of probable costs versus benefits. Nevertheless, a growing array of evidence suggesting the probability of failure may convince even the most politically oriented decision-maker.

Once it has been decided to investigate the feasibility of a particular kind of economic development affecting natural resources, expert integrated survey and evaluation should be fully called into play. The engineering or technological feasibility of the project usually receives careful attention with the result that few projects fail because of lack of adequate engineering —remarkably few dams collapse and highway bridges rarely break down from the weight of vehicular traffic.

When the engineering outlines of a proposed development are sufficiently advanced, economic evaluation is normally conducted to determine the costs of the project and their relation to the expected economic benefits. Such economic evaluation should take place before the technical plans for development are far advanced. This evaluation, which measures factors quantifiable in money terms, is often far less exact than the engineering studies; economic analysis often errs in the underestimation of costs and the overestimation of benefits (although sometimes it also errs in the other direction). In part such errors result from the normal uncertainties of economic existence—nobody can be certain of the world market for a crop five, ten or twenty years in the future. However, many errors also result from inability to quantify the complex environmental factors and values that will affect both short and long-term costs or benefits from a development project. For example, the ill-fated cross-Florida Barge Canal in the United States was originally planned and its route largely determined at a time when wilderness in Florida was abundant and undervalued. Years later, when construction was well under way, it was realized that the route would take the canal through one of the state's few remaining, and consequently now highly valued, wilderness areas. This environmental consideration was a factor that strongly determined the decision to abandon the project after the investment of 50 million dollars in the partially completed waterway.

Generally speaking, the array of expert advice normally brought to bear upon the prefeasibility analysis, preinvestment survey, evaluation and planning of a project is narrow and comprised of representatives of those disciplines that have traditionally been consulted in such matters. A decision to develop ground-water resources to provide livestock watering points in a semi-arid area will usually be based on the advice of competent geologists and engineers who will decide whether or not water can be made available, and at what cost. The ecology of the animal and pasture resource usually receives less careful attention. An animal husbandryman may be asked to evaluate pasture resources, or an expert in pasture management in humid lands may be consulted on problems of desert range management. Rarely will a range ecologist familiar with semi-arid pastures and livestock be brought into the picture. Other, less obvious problems such as those of water-borne disease may also be ignored by development planners. Even more rarely will a sociologist or anthropologist be asked to evaluate the traditions, attitudes and probable behaviour of the people to be affected by the proposed development. This does not imply that any of the usual types of expert advice should be neglected. On the contrary, politics, economics and engineering will always remain essential ingredients for development planning. However, the efficiency of those concerned with these traditional fields will be greatly enhanced, and the likelihood of reaching sound decisions will be improved, if the advice of those qualified in the environmental and social sciences is sought at an early stage in the planning process.

(4) THE SPECIAL ROLE OF ECOLOGY IN DEVELOPMENT PLANNING

Ecology is neither an emotional state of mind nor a political point of view, although there are those who seek to use ecological ideas to whip up emotions and influence politics. Ecology is a science. Like all science it is a body of knowledge and a means to attain further knowledge. This knowledge can be used for positive or negative ends. It can be put to work either in enhancing the human environment or in more effectively destroying it.

Ecology is the science concerned with relationships between living things and their environments. More usually it is concerned with the environmental relationships of populations and communities of living things. Because of the emphasis of this science, ecologists early became concerned with the ways in which all aspects of the environment interact with one another. It is impossible to study even a simple population of lichens growing on a rock without becoming involved in a study of the ways in which sunlight, water, the atmosphere and the earth's crust all affect these primitive plants and are in turn affected by them.

So long as ecologists concerned themselves with studies of biological systems in areas remote from human interference, their activities had little interest to those concerned with human affairs. But this condition did not for long prevail. During the 19th century, forest ecologists began to study the interactions in forests that resulted from human use and management of these areas. Early in the 20th century, rangeland ecologists began to study the interactions between man and his livestock and the plants and soil which supported them. In the 1930s, the need for conservation of wildlife and fisheries led some ecologists to study the effects of land and water management on fish and other animal populations.

Ecology is by necessity an integrating science, and must progress by bringing together the specialized knowledge acquired by physicists, chemists, hydrologists, meteorologists, pedologists and geologists, as well as many other relevant scientific fields. This knowledge is brought to bear on understanding the environmental relationships of the living things being studied. Ecologists must be, to some extent, generalists, but there is no all-purpose kind of scientist known as an ecologist who can be used to evaluate all the impacts of high dams on fisheries, of grazing livestock on shrublands, of logging practices on soil stability, and of urban development on human welfare. There is, however, an ecological discipline and way of working that may be instilled in a forester, agricultural specialist, range manager, wildlife scientist, public health specialist, landscape architect, development economist or engineer. It is this interdisciplinary point of view that has particular relevance to development planning. It will always involve bringing together the appropriate range of expertise to enable a full environmental evaluation of any proposed way for using land and water, or otherwise modifying the human environment.

The use of ecology in development planning has the aim both of enhancing the goals of development and of anticipating the effects of development activities on the natural resources and processes of the larger environment. Thus, an ecological appraisal of a particular type of development will focus on the area of resource to be exploited, and also on the larger environment and human inhabitants, whether of the region surrounding the site of the development project, or in the global biosphere as a whole. In the past, ecological knowledge has been employed mainly in assessing the potential productivity of a resource and in determining how it should be exploited. Relatively little regard for the environmental impact of a productive activity has been shown, other than concern for how such an impact might limit the attainment of production goals. However, certain kinds of development and management technology have had such adverse secondary impacts on local and global ecosystems that their effects must now be anticipated in the planning process and, to the extent possible, evaluated. An ecological approach to diagnosing such impacts will be necessary to understand their nature and significance. It is this application of ecology which is emphasized in the present book and which, it is hoped, will aid planners to make more sure of success.

Ecology cannot and should not attempt to provide the mathematical precision that characterizes physics and chemistry. In dealing with complex environments there are many variables and many possible outcomes from changes that may be made in these variables. Nevertheless, in numerous situations a high degree of predictability is possible, and where it is not, the probabilities of certain things happening can still usually be evaluated on the basis of past experience. There is no need to stretch ecological knowledge beyond its useful limits for it to prove valuable in economic development. For example, shortly after the pesticide DDT was discovered and used widely to control insects, ecological studies revealed its potential dangers. The probable consequences of its continued, widespread use were predicted long before they actually appeared in the field, and long before there was any great public interest in the subject. The precise way in which DDT affected the behaviour and reproduction of animals was not known until many years of field and laboratory work had been carried out, but the general effects of this pesticide were fully anticipated and these are discussed in the chapters that follow. Research on DDT could well have been considered as a notable example of the successful application of science if the ecological implications had been promptly accepted by all concerned with related aspects of development planning. It is unfortunate that they were not, because, as a result, environmental damage caused by DDT still continues and will grow even worse.

One function of ecological studies preceding development is therefore to predict damage that a particular action may inflict on the environment, thus enabling decisions to be made in full knowledge of probable consequences. A decision to go ahead may sometimes be justified by counterbalancing benefits that are sought, but it should never be taken blindly.

(5) THE PARTNERSHIP OF CONSERVATION AND DEVELOPMENT

Development of natural resources contributes to providing an increasing share of the necessities of life, and material luxuries of civilization, to the people of a country. Conservation of natural resources can assure that the environment resulting from development is one that will be satisfactory to the people involved and is self-sustaining or capable of being sustained, is healthful, challenging, and offers opportunities for future change. Ecological knowledge is as essential to conservation as it is to development. Many conservation efforts have failed and great amounts of money been wasted because ecological facts were not known or were ignored. The range of conservation values to be considered in development planning ought to be related therefore to ecological information, without which sentiment and subjective evaluation will prevail. Equally, however, the range of economic values for the same development planning must take account of the ecological information: otherwise there will be the same probability of errors of judgment.

For example, in a particular tract of land not yet opened to human use a wide range of options exists:

1. it can be left in a completely natural state and reserved for scientific study, educational use, watershed protection and for its contribution to landscape stability;
2. it can be developed as a national park or equivalent reserve, with the natural scene remaining largely undisturbed to serve as a setting for outdoor recreation and the attraction of tourism;
3. it can be used for limited harvest of its wild vegetation or animal life,

* *

but maintained for the most part in a wild state—serving to maintain landscape stability, support certain kinds of scientific or educational uses, provide for some recreation and tourism, and yield certain commodities from its wild populations;

4. it can be used for more intensive harvest of its wild products as in forest production, pasture production for domestic livestock, or intensive wildlife production. In this case its value as a 'wild' area for scientific study diminishes, but it gains usefulness for other kinds of scientific and educational uses; its value for tourism and outdoor recreation diminishes but is not necessarily lost; its role in landscape and watershed stability is changed, but may be maintained at a high level;

5. the wild vegetation and animal life having been removed in part, it can be intensively utilized for the cultivation of planted tree crops, pastures or farming crops; or

6. the wild vegetation and animal life having been almost completely removed, it can be used for intensive urban, industrial or transportation purposes.

So long as the first three choices are taken, the option remains open to change from one of these uses to the other or to use the land for any of the latter three purposes. If the fourth choice is picked, the options for restoring the land to any of the first three categories are reduced but not eliminated. Selection of the latter two development possibilities largely prohibits any shift to the other alternatives within a reasonable period of time.

A rational and sensible choice from among the options available must be based on ecological and economic considerations, as well as on other grounds. Some areas have extremely high environmental values—they may represent unique communities or support rare or endangered species, or they may be essential to maintain soil stability and water yield in a river basin. Where alternative areas exist that are suitable for more intensive uses, it would be absurd to use an environmentally unique area for any purpose liable to damage its environmental values. Some areas, however, have extremely high economic potential. They may represent a unique source of valuable minerals; provide the only logical site for a hydroelectric dam and reservoir; support a high-yielding, high-value stand of commercial timber; or have deep, rich soils which could be put under intensive agriculture for a long time. Such areas are, of course, logical sites for the more far-reaching forms of development. If, by bad fortune, high economic development values and high values for ecological protection should coincide on the same tract of ground, all of the skills of both economics and ecology may be required to find ways to maximize the total gains to the community and minimize the losses.

In most developing regions and nations numerous conservation and development options remain open. It is always easier to take into account conservation needs before intensive development begins. It is far less expensive to protect than to restore the environment, and some types of environmental damage are irreparable. As a first general principle it can therefore be stated that:

In making a decision to develop hitherto untouched land, the need to keep a range of resource use options available to future generations should be a major consideration.

(6) TIME SCALES

Political and economic factors in determining resource uses tend to be particularly influential in the short run. The politician who does not produce observable results is not likely to stay in office and may lose the opportunity to participate in directing the nation's development. The economist who forecasts demands, supply, and probable prices too far ahead is almost certain to look foolish when that time arrives.

Economists view the preference for long-term results as redistributing income from present to future generations. As one economist (Gordon Tullock, 1964) put it—

'Are there so few diseased, illiterate, underprivileged today, so few persons who excite our sympathy that we must look to the prospectively wealthy future for a source of worthy recipients of our bounty?'

His implication is clear: resource economics has stressed immediate, short term use and de-emphasized long-term, less quantifiable social needs. This view has often advocated cutting the trees now, mining the ore now—and at the lowest possible cost. At the same time it advocates not diverting capital from the public coffers for less immediately productive purposes, such as open space purchase and protection, pollution control, improvement of nutrition and public health, and preservation of organic diversity.

By contrast, the ecologist is often concerned with the long-term effects of factors that may show little evidence of operating at all in the short run. Similarly the conservationist is persistent in his emphasis on long time scales; he would be inclined to question the implications of Tullock's remarks on the needs of today's people. If carried to an extreme such an emphasis could destroy any hope for a 'prospective wealthy future' through exhaustion of those resources on which future generations must depend. To quote the United Nations Advisory Committee on the Application of Science and Technology (1970):

'Every operation to develop a natural resource has both short-term and long-term results, and an attempt must be made to anticipate these results so that the choices that have to be made may be based on the fullest possible knowledge of the problem. In reaching decisions, the short-term results are likely to be given more weight than the long-term results, but care should be taken to ensure that the latter should not lead to an irreversible deterioration of the resource that is being developed.'

Attention to this advice should minimize conflict between those concerned with economic development and those concerned with conservation values. Nevertheless hard decisions must still be faced. For example, heavy exploitation of a surface mineral resource may virtually destroy for a long period of time the other values of an area of land, but without such exploitation a nation may be unable to accumulate sufficient funds to carry out needed conservation or development activities in other areas. Where such is the case, a reasonable concern for the future would dictate that a portion of the capital generated by the development be expended towards the rehabilitation of the area after mineral exploitation is completed. Further, where mineral resources are extensive it may well be in the national interest to leave untouched certain areas of high mineral value while developing other areas, thus retaining a wider range of options for the future and protecting for the longest period of time other important resources of the mineralized region.

(7) ALTERNATIVE CHOICES

Development goals can ordinarily be reached by a variety of different paths. Some of the paths lead to irreconcilable conflicts between those concerned with the environment and those concerned with commodity production. If either side becomes too fixed in its position, conflict results and often both sides lose. Nearly always an alternative, more compatible choice can be found. In the case of the proposed international jetport in southern Florida, transportation experts, economists, engineers and politicians reached a decision to locate the jetport in the Big Cypress Swamp at the edge of Everglades National Park. Ecologists and conservationists were not consulted with reference to this decision and were actively discouraged from intervening. As a result, a long legal and political battle resulted in which the jetport planners were ultimately instructed to find a new site for their facility. By then they had invested 13 million dollars in the Everglades/Big Cypress location. Alternative sites had always been available and could have been given early priority if ecological advice had been sought from the beginning.

In most regions, existing patterns of land use commonly reflect hundreds or thousands of years of human trial and error tested against the ecological limitations governing the environment. Commonly the most productive sites are those already intensively settled and occupied. The remaining wilderness and wild land is usually that which has been least suitable for human settlement. Admittedly, changing technology alters both human capability and, to some degree, environmental restrictions. But in general, and with some notable exceptions, it can be stated as a principle that:

Greater returns for investment in agricultural and pastoral development are obtained from enhancing or increasing the yield from already developed lands of known and proved productivity than from the attempt to convert still wild and, therefore, probably marginal lands to some highly productive form of commodity use.

Concentration of development planning on proven high-quality lands can and should take advantage of past experience, through careful and explicit studies; it will also decrease the likelihood of conflict with the important objective of conserving some more marginal areas for other purposes.

Wild species cannot be restored once they become extinct. Unique landscapes containing varied and interesting natural communities can rarely be put back together once they are greatly modified by human use. In any nation these are irreplaceable resources whose value can only be measured over the long run through the long-term satisfaction they give to human needs and aspirations, and their long-term contribution to the stability and productivity of the land. *Protection of species and natural communities is a logical first step in the development of a region.* When this has been provided for, other areas can be planned and utilized for a variety of other more intensive human purposes. With reasonable care and management intensively developed areas can produce the commodities that man requires on a sustaining basis. The ecological concepts in later sections are related to the processes relevant to such careful management.

It is no longer necessary to make major blunders in land use through lack of ecological knowledge. It is no longer necessary to sacrifice long-term environmental values for short-term economic needs. It is possible to reconcile the conflicting demands upon lands and resources so that irreparable damage is avoided and human populations can enjoy meaningful benefits.

(8) REFERENCES

Tullock, G., 1964. The social rate of discount and the optimal rate of investment. *Comment, The Quarterly Journal of Economics,* Vol. LXXVIII, No. 2, pp. 331–336.

United Nations Department of Economic and Social Affairs, 1970. Natural resources of developing countries: investigation, development and rational utilization. *Report of the Advisory Committee on the Application of Science and Technology to Development.* UN, New York, 174 pp.

CHAPTER 2

General Ecological Considerations

Some concepts developed from the science of ecology have general application to all areas and to all situations in which economic development may be expected to occur. These will be the principal concern of the present chapter; their application to some particular ecosystems and particular types of development of major importance at the present time, will be further elucidated in the remaining five chapters of the book. From very early times man has advanced his own ends by the modification of natural communities and ecosystems, sometimes in minor ways to enhance the yield of a desired commodity, but with the rise of agriculture, civilization and modern technology in ever more drastic ways. Such modifications are most likely to be productive and stable when the environmental limitations of the natural ecosystem are well understood. In the following sections, therefore, the structure and functioning of ecosystems are examined, including their relation to the biosphere. All life depends on how successfully man can learn to harmonize development with these environmental systems.

(1) ECOSYSTEMS AND THEIR FUNCTIONING

Any assemblage of plants and animals able to exist within an area will in time form a biotic community. In this community the different species tend to interact with one another and to modify the conditions of life within which each exists. They therefore develop the interrelationships and interdependences which constitute an ecosystem.

In a forest grove or woodland, for example, the larger trees—the dominants—influence the environment in which all other species live. Plants which cannot grow in their shade or tolerate the competition for soil, water and minerals provided by their root systems, or the various chemical substances that they may release into the soil, are unable to find a place within the community dominated by these large trees, or will be confined to places where such trees cannot maintain themselves. Thus, the trees exert an obvious influence on other plants. However, smaller plants may modify the environment in various ways to enhance or detract from its capacity to support the larger trees. The inability of various pine trees to thrive in the absence of the mycorrhizal fungus that attaches itself to their roots has long been known. Similarly, legumes' roots provide a home for bacteria which in turn manufacture the nitrates on which legumes depend. The animal components of such a biotic community also modify the conditions of life for the plant (and are, in turn, affected by the plant community). The interrelationship (symbiosis) between the *Yucca* plant and the yucca moth is a well studied example; the plant supplies food for the moth, but is dependent on the moth for fertilization and perpetuation. Similarly, large grazing animals affect the populations of the plants on which they feed, and are dependent upon the quality and quantity of plant growth for survival. Carnivores may also limit or otherwise modify the populations and the behaviour of herbivores, and thus have a secondary impact on the kind and intensity of grazing.

It is convenient to draw a distinction here between 'natural' and 'artificial' communities. The former consist of wild, naturally occurring species of plants and animals able to maintain themselves in the absence of man. Artificial communities are characterized by species introduced by man or favoured by human modification of the environment, and are unable to exist without continued human assistance or interference.

Neither the word 'natural' nor 'artificial' is entirely satisfactory, since man is part of nature, and all communities, whether strongly influenced by man or not, are also a part of nature. Furthermore, it is difficult to find any area of the earth in which some human modification of the environment has not occurred, if no more than a minor change caused by an increased frequency of certain radio isotopes, or chemical fall-out from air pollution.

Both categories, though useful to distinguish for some purposes, always exist as part of *an ecosystem,* which *is essentially a biotic community in interaction with its physical environment.*

The physical environment is composed of sunlight, atmosphere, water and soil or rock. All of these physical elements affect the living organisms, and all are in turn affected by them. Plants, for example, obtain their energy from sunlight, their basic food from air and water, supplemented by chemicals from the soil or rocks. Plants, however, differentially reflect, refract or absorb various wavelengths of sunlight, and thus modify it; by using carbon dioxide and by giving off oxygen and water vapour they modify the atmosphere; by removing various chemicals and by adding others they modify the soil. All plants and animals always exist as part of an ecosystem. Man, like other animals, is dependent upon the ecosystems in which he exists despite his high mobility which enables him to move from one to another (or to extract products from one ecosystem and export them to another), and despite the technology which permits him to create major modifications in any ecosystem.

All ecosystems must have certain component parts or functions. Thus, they must have a source of energy (usually sunlight). They must have organisms capable of converting this energy into chemical or food energy (usually green plants). They must also have organisms capable of taking the chemicals built up by green plants and breaking them down into simpler forms in which they can be re-used. These include the so-called *reducer* organisms, such as bacteria and fungi, that act in the breakdown and decay of dead plants and animals. The living components of ecosystems are the plant and animal species (including micro-organisms) that make up the biotic community. Each species is adapted to a particular role in the ecosystem known as its *ecological niche.* Each depends for its existence upon the presence of a suitable habitat, comprising other species as well as necessary components of the physical environment. The functioning of the ecosystem is dependent upon the presence of a suitable combination of species each of which performs a specialized task within the total system.

In a natural ecosystem sun energy will be captured by trees, grasses, or other herbs, and converted into proteins, carbohydrates and other food components. These will be consumed by plant-eating mammals, birds, insects and a great variety of other animals, which, in turn, will provide food for meat-eating animals. All ultimately provide food for the organisms of decay which will break down plant and animal materials and restore their chemical components to the soil.

Energy will pass through an ecosystem in a one-way path. Of the solar energy reaching the earth, only a portion will be stored by plants; often less than 1 per cent of the total sunlight energy falling on a vegetated area is retained as chemical energy by the green plants. Of the total energy thus available in plant tissues for consumption by animals, usually less than 20 per cent will be stored as chemical energy in an animal's body tissues.

Eventually, after supporting life for a time, most of the energy is lost to the ecosystem. Some, however, is stored in long-lived organic materials such as tree trunks, or in dead organic materials preserved from decomposition—those that eventually form organic deposits of one kind or another, including peat and, over longer periods of time, oil, coal and natural gas.

By contrast, chemical nutrients flow through an ecosystem in circular pathways, from soil to plants and animals and back to soil, being reused or recycled again and again.

(2) THE BIOSPHERE AND ITS FUNCTIONING

All of the ecosystems of the earth, together, form the biosphere. The biosphere is that portion of the earth within which life exists. It includes all oceans and freshwater, the lower layers of the atmosphere and the outer skin of the earth's crust—the rocks and soil of the earth's surface.

Within the biosphere all of the functions and structure described for any one ecosystem exist. Green plants capture sun energy and combine it with chemical raw materials from soil, water and air. The food they produce supports all animal life, including the decay organisms which return it to the soil for plant use once more.

Man is a part of the biosphere and depends on its continued functioning for his own existence. He can modify or even destroy any one ecosystem. He cannot, however, risk major modifications of the biosphere except at the risk of his own extermination. Thus, the continued production of plant materials, whether wild or cultivated, is the basis for the nutritional support of man as well as all other animals. The continued functioning of green plants is the source of atmospheric oxygen on which man and other animals depend. The continued functioning of the reducer organisms is the means by which the chemicals in human wastes or in the bodies of plants and animals are made available for further use by living things. A breakdown in any of these biospheric systems would imperil human survival.

Man has been able to take major risks in modifying local ecosystems. He cannot afford major risks in dealing with the functioning of the biosphere.

(3) FACTORS INFLUENCING POPULATION GROWTH

The growth of any species population, whether it be of trees, wild animals or men, depends upon the excess of births (natality) over deaths (mortality). For any local area or ecosystem this may be influenced by immigration, movement into, or emigration, movement out of, the area or system concerned. For the biosphere as a whole, however, there is no emigration or immigration.

Populations, unless otherwise checked, tend to grow rapidly when there is an abundance of space and of the materials that the species requires for its subsistence. They grow much more slowly when space or essential materials are in short supply. They level off or decline when space or materials become limiting.

A population growing at the maximum rate for which that species is capable is said to be growing at its biotic potential rate. Such a rate of growth is only exhibited when conditions are most favourable for the species —when natality is at a maximum and mortality at a minimum. Operating to prevent a species from maintaining a biotic potential rate of increase is the environmental resistance. This is the sum total of all of those factors that cause mortality or decrease natality. When biotic potential equals environmental resistance, the species population will not grow.

Most animal populations have long been established in the area which they occupy. They seldom show either a continued growth or decline unless the conditions of their environment change drastically. Most show a fluctuation around some mean level which is determined by the ability of their environment to support their numbers.

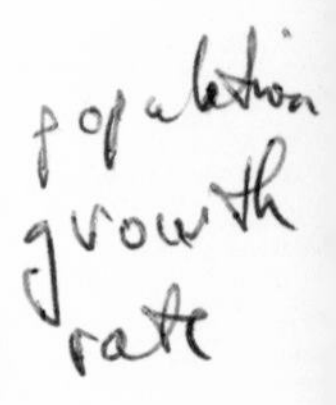

Populations of animals (or of men) introduced into an area previously uninhabited by the species concerned but otherwise suitable for it, tend to show an increase in a manner which resembles that of an S-shaped or logistic curve. This is at first marked by a rapid (exponential) rate of increase, and then by a period of levelling off towards some relative stability. Previously established populations, when freed from some factor of environmental resistance, may show a similar growth pattern. There are many modifications of the shape of the curve, depending on the species and the conditions of the environment, but the eventual period of levelling off, through increased environmental resistance, is characteristic of all growth curves.

(4) LIMITS TO POPULATION GROWTH

All populations must sooner or later level off their curve of growth. This levelling-off can result from the behaviour of an animal species. Some species are intolerant of crowding and produce relatively fewer young as their habitat becomes more fully stocked. Other species, however, are more directly limited by the action of factors outside of the population which cause increased mortality. Thus population growth must inevitably be limited by the environment. No environment is without limits and the biosphere is limited both in extent and resources.

If there were no other restrictive factors, ultimate limits to the increase of any species would be imposed by the amount of sunlight energy reaching the earth. In fact, however, other factors are always limiting. For plants, limits to growth would be set primarily by the extent of their ability to convert sunlight into food energy, or by the availability of carbon dioxide in the atmosphere. In the operation of natural ecosystems, however, either water, or chemical nutrients present in soil or water, run out before the limits of carbon dioxide or photosynthetic efficiency are reached. Animal populations must ultimately be limited by the availability of the food that they eat, which depends on the ability of plants to continue supporting herbivorous animals. In practice, most animal populations tend to be limited by other factors before the absolute limits of food supply are reached. These other factors are most notably behavioural, those responses of animals to the presence of other members of their species that prevent an excessive accumulation of individuals in any one area. Interspecific predation can also be a limiting factor of some importance.

Figure 1. Human populations cannot escape the limitations on growth imposed by the resources of the biosphere. (Photograph by Ringier Bilderdienst: courtesy World Wildlife Fund).

Human populations have attained comparative freedom from the limitations imposed by any one ecosystem—the exceptions being the remaining primitive groups of people. However, human populations cannot escape the limitations imposed by the resources of the biosphere (Figure 1). Of particular relevance is man's efficiency in converting those resources for use and in recycling waste products or by-products for ultimate reuse.

(5) CARRYING CAPACITY

The environmental limits to the growth of any one species population determines the carrying capacity for that species. Carrying capacity is a measure of the number of individuals of any species that a particular environment can support.

Carrying capacity can be considered as having several levels:

1. an absolute or maximum carrying capacity which is the maximum number of individuals that can be supported by the resources of the environment at a subsistence level (this level can be termed the *subsistence density* for that species);
2. the level at which a species population is normally held by the influence of other species living in the same environment—those that hunt or prey upon the species, and those that cause disease or parasitic infestation (this level has been termed a *security density*, or threshold of security, because populations below this threshold are relatively secure from predation or disease; this is necessarily less than the subsistence density);
3. a level, which is generally considered more desirable by those concerned with the health or productivity of the species involved, termed an *optimum density*. At this level individuals in the population will have available an adequate supply of all essentials for existence and, in consequence, will show abundant individual growth and health not limited by shortages of any essential requirements. Such an optimum density can only be maintained by strong limitations on growth imposed by the behaviour of the species concerned (self-limitation) or by removal of individuals in excess of this density by the action of other species through predation. In the latter case optimum density and security density will be the same. These are not necessarily fixed levels, but will fluctuate upwards or downwards.

Intelligent management of domestic animals for maximum production of meat, milk, etc., involves holding their populations at or near an *optimum density*. At this level, their productivity will be at a maximum and the highest yield can be obtained. Under primitive conditions in Africa and Asia, for example, domestic herds were often held at or near a *security density* by the action of predators (human or wild animal), or by diseases and parasites. Under modern conditions, with predation and disease minimized, and where they are not properly managed, domestic herds often increase to a *subsistence density*. Under these conditions malnutrition is common, disease is more prevalent and productivity is low (Figure 2).

Obviously, man has similar choices with his own populations. In primitive times the operation of predation, disease, etc., often held local populations at or near a security density. With the removal of predators, and the control of disease, populations can either increase to a subsistence density, a level where human suffering and general misery will be maximum, or they

Figure 2. At a subsistence density livestock malnutrition is common, disease is more prevalent and productivity is low. (Photograph by Kenya Information Service).

can be levelled off by self-imposed restrictions on growth, at or near an optimum density. This would allow maximum individual opportunity for health, happiness, and choice of high quality life styles and environments. These choices would normally be available to all human populations at some period in their growth. Now for the first time, however, the whole human species faces possible population increase to levels in excess of the capacity of the biosphere to maintain them, the biosphere itself being strained by rising demands for material resources and products.

Carrying capacity is not fixed. It fluctuates naturally with weather and climate, and the operation of other natural factors such as fire, floods, earthquakes and vulcanism. It is being modified continually by human action. In any one area it can be increased to some maximum level which is determined by the rate at which that environmental requirement which is in shortest supply can be provided. Such a requirement is known as a limiting factor. A limiting factor is that substance or quality in the environment, the supply of which is least abundant in relation to the needs of the animal or plant concerned.

(6) LIMITING FACTORS AND THEIR OPERATION

Limiting factors are often classified in two categories: density dependent and density independent. The effect of the former on a population increases as the population increases in density. It is essentially a stabilizing influence on growth and causes a population to level off at carrying capacity. A density-independent factor, conversely, affects many or a few individuals without reference to the population level.

Food supply is generally density dependent—the more there are to eat it, the less there is for each individual, and the greater the effect of food scarcity. By contrast, a flood is density independent since it may wipe out an entire population of a species, whether there are few or many. The more important limiting factors are as follows.

(a) Climatic and Atmospheric Factors

These operate in a variety of ways to affect species populations. To begin with, any species has limits of tolerance and some optimum range of tolerance for such factors as sunlight, temperature, humidity, rainfall or wind. If a local climate normally exceeds the limits of tolerance for a species, then the species will not occur in that area; if it exceeds the optimum limits, the species will not thrive there. If the local climate only rarely exceeds the limits of tolerance, then a species may temporarily occupy that area, but will be eliminated in those years when the climate becomes extreme.

Assuming that the climate is favourable to the establishment of a species, then a number of climatic factors may influence population growth. Changes in temperature—years that are warmer or colder than normal—may permit a species to thrive and increase, or they may cause a decrease and permit its survival only in the most favourable sites. Changes in rainfall and humidity, dry years or cycles and wet years or cycles, have major effects. Fluctuations in temperature and rainfall tend to be most severe in areas where temperature or moisture is already near the limits of tolerance for a species. Thus, in the humid tropical lowlands where rainfall and temperature are high and near the optimum for the growth of the greatest number of plant species, there is relatively little annual fluctuation in either factor. In the dry tropics not only is moisture naturally limiting for plants, but fluctuations in rainfall from year to year may be extreme. In northern temperate regions where winter·cold and the duration of low temperatures restrict the growing season for plants, one finds the greatest variation between temperatures in one year and another.

Climatic and atmospheric factors tend to have their most severe effects upon those introduced populations of plants and animals which man attempts to grow or produce in areas that have climates more extreme than the optimal climate for the species concerned.

Native species are adapted to the range of climate in the area they occupy, and although they may be affected by climatic extremes, these effects are usually much less limiting than for a domesticated species introduced outside its normal climatic range. In the African savannas, a drought will be more likely to wipe out a high percentage of the cattle than of the native antelopes. Climatic and atmospheric factors also have more effect on species that are at, or approaching, the level of subsistence density. Well-situated animals or plants, with adequate nutrients, water and other essentials, are less likely to be affected by drought than populations which have increased and expanded into marginal areas and are already in poor condition because of nutritional limitations. Certain factors of weather and climate may have catastrophic effects on all populations in an area—hurricanes, tornadoes, unusually severe floods or droughts, but usually these are less significant than the more normally occurring changes, except in the case of those species which are highly limited in distribution. With most species, normal fluctuations in climate have their greatest effect when animal population densities are excessive, or when plants are established in sub-optimum environments.

(b) Soils

Here the limiting factors operate directly on plants and, through plants, upon animal populations (other than those of the soil fauna itself, which is often directly limited by soil conditions). The requirements of plants and animals, however, are not identical.

An abundant growth of vegetation does not necessarily mean there will be an abundant food supply for animals, nor that the food supply will have high nutritional value. Certain plants may produce abundant vegetative growth and high amounts of carbohydrates under conditions where certain soil chemicals are in short supply; however, they may produce relatively small amounts of the proteins and vitamins needed to support an abundance of animal life.

Thus, a shortage of a trace element, such as cobalt, in the soil may have little effect on plant growth. However, cattle feeding on these plants may do poorly or not survive. Addition of a cobalt top-dressing to the soil which is then taken up by the plants can open up a rangeland to livestock use. For another example, a dense forest may produce an abundance of vegetation, but will support very few ground-dwelling grazing or browsing animals. There is no lack of food for them, but plants growing on a forest floor under dense tree cover are commonly lacking in protein. Animals grazing or browsing them can receive adequate carbohydrates but not enough protein to thrive or reproduce. Thus, in many areas of dense forest, grazing and browsing animals will only be common around natural or man-made clearings.

The abundance of natural vegetation in an area often has little relation to the inherent fertility of a soil. Thus, the ferralsols and acrisols* of the humid tropics are notoriously infertile, yet they support the most luxuriant vegetation to be found on earth. Examination of this situation reveals, however, that virtually all the nutritional elements required for the support of life are tied up in the vegetation and animal life of the area, and recirculate quickly from the dead plant or animal back into a new, living individual. Below the organic surface layer the soil contains few of the elements needed for plant or animal nutrition, and if the vegetation is scraped away with a bulldozer the soil that remains is often highly infertile. By contrast, some desert soils that support little plant life, because of lack of water, may have a rich supply of the soluble mineral elements needed for the support of life. When water can be made available, they will support high levels of productivity, although there may be new ill effects such as salinization, waterlogging and water-borne diseases.

(c) Water

Water is an obvious limiting factor for plants and animals. Plants are classified in relation to their tolerance for dry, medium or wet conditions as *xerophytes, mesophytes* and *hydrophytes*. Most land plants are in the intermediate, mesophytic range; desert plants are usually, but not always, xerophytes; hydrophytes grow in water or require an abundance of water in the soil. All plants require water to support their active growth and metabolism. Some plants cannot tolerate even brief periods of moisture deficiency and will wilt and die when the soil dries out. Xerophytes have special adaptations that permit them to survive under prolonged drought, but must have some moisture in order to grow and reproduce.

Where present in abundance, water limits the occurrence of terrestrial organisms and provides a habitat for aquatic forms of life. The latter will relate in their diversity and abundance to the quality of the water, its ability to supply oxygen, its salinity, temperature, velocity and other factors.

* Terminology from FAO/UNESCO World Soil Map. See Dudal (1968).

All of these are capable of being affected, usually adversely, by pollution or siltation.

In relation to terrestrial vegetation, an abundance of water on or in the soil limits plant growth primarily by restricting the availability of oxygen. Those plants that grow in wet soils or flooded areas have the ability to derive oxygen through special structures, such as devices that permit the aeration of their roots; the pneumatophores of mangroves, or the so-called 'knees' of bald cypresses are good examples. Continuously moist or flooded soils inhibit the bacterial or fungal action that would otherwise bring about decay of organic debris and litter. Organic remains will usually accumulate on such soils in the form of organic mucks or peat.

Animals, like plants, require moisture for their continued existence, but vary greatly in their ability to endure drought. Apart from purely aquatic animals, amphibious animals such as the crocodile and hippopotamus cannot long survive away from water. Most land animals must drink water regularly in order to continue to live. However, some desert species can exist indefinitely in the absence of free water as long as the food which is eaten contains sufficient moisture and they can find enough shelter to hold the loss of water from their body surfaces to a minimum. A severe drought, bringing an absolute shortage of water to an extensive area, can be fatal to all animals that require water for drinking. Nevertheless, much of the mortality attributed to drought results not from a shortage of water, but from a shortage of food within reach of water.

(d) Biotic Factors

The greatest number of limiting factors influencing plant or animal growth, abundance and distribution are biotic in nature. Food supply for animals is one of these, and is the most common factor limiting the growth of animal populations, either directly, through being short of requirements, or indirectly, through behavioural responses to food shortage. The number of plant-eating animals in any area is ultimately limited by the abundance of the plants on which the animals feed. Furthermore, since the plants must survive and reproduce, there must always be a greater abundance of plants than what is actually needed to provide food to the animals. Most plants produce a surplus of vegetative growth and of seed, part of which can be used by animals. However, each plant, in order to survive, must maintain a metabolic reserve—a minimum amount of leafage to permit it to store food for its own survival, or set seed for its own reproduction.

Plant growth may be limited by competition from other plants of the same or different species each drawing on the same reserves of soil and water, or shading one another from essential sunlight. Some plants secrete substances which inhibit the growth or establishment of other plants. A variety of organisms may prey upon plants, from the seed stage through the life cycle to the mature plant.

It is obvious that the number of carnivores or meat-eating animals in any area is limited by the availability of the prey upon which they feed. Also, there must be enough prey animals to enable the prey species to survive and reproduce; the predator cannot eat all of the prey, or both would become extinct. Most prey animals have developed behaviour patterns in relation to their habitat conditions that permit some of them to escape predators. Predators seek the most vulnerable and available prey, and usually do not hunt down and capture the more elusive individuals. Often the individuals captured are either the very young and very old, or the most diseased or genetically least vigorous stock; therefore, natural predation tends to have overall beneficial effects on the population of a prey species.

The relationship of the abundance of predators and prey is sometimes illustrated in the form of a *biotic pyramid*. Green plants occupy the base of the pyramid and must always have a greater total volume, or *biomass* than the plant-eaters that feed upon them. The latter, the herbivores, form the second step in the pyramid, which must in turn be wider and represent a greater biomass than the third step, representing the carnivores that feed upon herbivores. Furthermore, if there are carnivores that eat other carnivores they must be fewer in number and contain less biomass than their prey. Each succeeding level in the biotic pyramid also represents a consistent loss of energy from green plants to herbivores to predators, since at each transfer of energy from one level to another some energy is necessarily lost. Far more calories are available at the lowest level of the pyramid than at any higher level.

The sequence of plants and animals feeding upon one another is called a *food chain*. Food chains are interconnected, since more than one herbivore may feed on a plant and more than one kind of carnivore may eat a herbivore. These interconnected chains form a *food web*.

As elements pass from soil (or atmosphere) to plants, and are then eaten by herbivores and in turn by carnivores, they are often concentrated in varying degrees by a process known as *biological magnification*. Thus, the element iodine may be present in the soil in small quantities and taken up by plants as it goes into solution and reaches plant roots. It has little direct function in plants, but is essential to plant-eating mammals. These must obtain enough iodine from the plants they eat and concentrate it in their thyroid glands for these glands to function properly. The level of iodine in an animal's thyroid is far greater, therefore, than in either the plants or the soil.

This well known situation achieved prominence when it was discovered that radio-active iodine was released by nuclear explosions and in certain wastes from nuclear plants. Although present at a very low and presumably safe level in the environment, radio-active iodine can become concentrated to a potentially dangerous and harmful level in the thyroids of animals. Similarly, other toxic chemicals that enter food chains may become concentrated as they pass up the chain, and will usually have their greatest effect upon the terminal predator at the end of the chain. This may be man himself, but it can equally well be a fish-eating bird, a mouse-eating hawk, or a rabbit-catching wild cat (see Figure 3).

Figure 3. Osprey, top carnivore on a food chain. Through the process of biological magnification, toxic substances, such as DDT, are concentrated as they pass from plankton to fish to ospreys. Even though their general level in the environment is low they may accumulate to a harmful level in species such as this. (Photograph by Eric Hosking).

Although the number of herbivores is limited by the supply of plants, and the number of carnivores is determined by the numbers of their prey, the reverse is also true in varying degrees. The abundance and distribution of a plant can be limited by the presence of herbivores. If a plant is introduced into an area where there are no species that feed upon it, it can spread rapidly and become transformed into a weed or pest. St. John's Wort or Klamath weed, when introduced into Australia and later into the north-western United States, rapidly became an important range pest. It was only controlled when one of the insects that normally feed upon it was also introduced. Similarly, the abundance and distribution of plant-eating insects and other animals is often controlled by a wide variety of diseases, insects or other predators that feed upon them. If the predators are destroyed, their plant-eating prey can increase suddenly and explosively, since its reproductive rate in nature is adjusted to the need for survival under constant predation. Sudden increases in numbers of rodents or other plant-eating mammals have often been related to the extermination of the predators that formerly held them in check.

Parasite and disease organisms play the role of minor predators. Frequently they have a direct, density-dependent relationship to the hosts upon which they feed. When the host animal is abundant, and particularly when it is weakened by shortage of food, its parasites and diseases will often become more widespread and have more serious effects. Most animals are adapted to survive in the presence of a wide range of diseases and parasites, and these have relatively small effect upon them providing that other conditions in their environment are favourable. Animals introduced into a new environment, however, are particularly vulnerable to the diseases and parasites of that area to which the native animals may have a high resistance or immunity. Similarly, diseases introduced into a new area can have devastating effects upon native species which have no immunity to them. Thus rinderpest, introduced into eastern and southern Africa by cattle in the 19th century, spread into wild game populations and caused serious reductions, almost to the point of extermination, among many species. Myxomatosis, an introduced disease, nearly wiped out the over-abundant population of European rabbits in Australia, although for the past ten years the percentage of resistent rabbits has increased with each eruption of infection.

Thus, it is characteristic of even the most virulent introduced disease organisms and parasites that they do not wipe out all the individuals of a species. In a genetically variable species, some individuals will possess a greater resistance or immunity, and some of these will survive. If the immunity or resistance is heritable, these survivors may breed up a new parasite- or disease-resistant population.

This general statement applies as well to plants as to animals. Plants are adapted to the diseases and parasites present within the ecosystems in which they normally occur. If introduced elsewhere they may be vulnerable to attack by a wide range of organisms from which they have no protection. A wild species of plant, because of its genetic variability, is more likely to survive attack by a new disease or insect pest than is a genetically uniform, cultivated variety.

Populations of many animal species are limited by the effects of interactions among individuals within a species. The species concerned are described as *territorial* and comprise a proportion of individuals which persistently or seasonally occupy a particular area to the exclusion of other individuals of the same species. Commonly, territorial animals will tolerate or encourage the presence of certain other members of their species—a mate or mates, their juvenile offspring, and sometimes the members of a larger social group, a herd, troop, flock or pack. However, they will drive off or otherwise exclude animals that do not belong to the tolerated category. Territorialism tends to limit the number of animals of any one species that occupy an area and to reserve for the territorial individual or group an adequate habitat and food supply; however, there are many kinds and degrees of territorial behaviour, and not all of them serve directly to limit population increase. Plants might also be considered to compete for 'territories'; the well-established individual holds its ground against competitors. Their territorial defence mechanisms include the shading out of competitors and the secretion of chemicals which prevent rivals from germinating or growing, already mentioned as limiting factors at the beginning of this subsection.

Numerous studies have now been made of the effect of overcrowding on a variety of animal species under conditions where food supply is abundant, and all other essentials (except space) are present in adequate supply. All of these studies show adverse effects upon individuals in a population where numbers exceed a certain level. These adverse effects are caused by social interactions, and vary from mild behavioural disturbances to serious forms of behavioural pathology, leading to increased mortality and a decline in birth rate. In some instances, widespread mortality is apparently brought about by a breakdown in the functioning of the endocrine glands.

(e) Interaction of Factors

Virtually all limiting factors interact and either reinforce or diminish their mutual effects. Thus the effects of animal overcrowding may be seen in the limitations imposed by diminished food supply, increased predation, greater mortality from disease and various behavioural disturbances. Similarly among plants, factors of soil, water, air, nutrients, the presence or absence of intense sunlight, all interact to influence the germination or growth of a plant.

Simple solutions to biological problems are therefore seldom likely to produce permanent, positive results. Modifying the impact of one limiting factor may simply increase the operation of others. For example, in an area of scarce water supply it may seem sensible to make available additional water in order to permit animal populations to increase. Yet doing so has, in many observed instances, quickly led to major die-offs from malnutrition resulting from an insufficient food-supply for the increased population. Similarly, destruction of predators may cause protected herbivores to die in an epidemic and the elimination of a disease may cause a crash die-off from starvation. It is therefore always necessary in management planning to look at the operation of the total ecosystem and consider how all the factors interact.

Failure to recognize these principles has caused many economic development activities to fail and much money to be wasted.

(7) DIVERSITY VERSUS SIMPLICITY

Diversity, in this context, refers to the number of species of plants or animals that live in a particular area. It has long been recognized that biotic communities found in more extreme climatic (and geological) environments are relatively simplified and less diverse than those occurring in climatic regions more consistently favourable to the growth and survival of living things. Thus, the cold subpolar regions of the world support relatively few species of plants and animals, the same being true of arid deserts, high mountains and the slopes of recently active volcanoes. The greatest diversity of plant and animal species is found in the humid tropical regions of the earth, where climate is relatively stable, uniform and favourable to plant life. In the more simplified natural communities, the lack of species variety is matched by the frequent occurrence of large numbers of a single species: the great herds of caribou or reindeer in undisturbed areas of the Arctic, or the enormous colonies of penguins in the Antarctic are good examples. The extensive stands of spruce in the sub-Arctic taiga illustrate the same principle. By contrast, such relatively high numbers of a single species are not commonly observed in the humid tropics, except where special edaphic, physiographic or successional conditions prevail, as in the case of mangrove belts in estuarine areas. Indeed the rarity of big stands of a single valuable timber species has been a factor militating against long-term economic management of tropical rainforests. In less diversified natural communities there are also great fluctuations in populations of some animal species and annual plants from year to year. The cyclic increases and declines in the numbers of arctic lemmings are a classic example. Similar fluctuations occur in many other arctic species. In desert regions, population fluctuations are not as regular, but may be even more severe; examples are the locust plagues and the massive increases in numbers, followed by major die-offs, of rodents or hares.

It has been concluded by many ecologists that diverse natural communities, because of the more numerous interactions among species, are more stable than simple communities. Furthermore, such communities are more productive in a gross, biological sense although the gains in organic matter production are generally off-set by losses through increased respiration. When these communities are simplified by human intervention, energy is channelled along more limited pathways and the species that remain may grow rapidly and achieve high levels of net productivity. However, because the normal buffering actions, or checks and balances, present in diverse communities have been removed, such communities are comparatively unstable. High net gains in organic production by one species in one year may be balanced by major losses in another.

Management of the human environment has generally taken the form of simplifying complex biotic communities, removing those elements less immediately useful to man, and substituting species more obviously useful. A complex rainforest is reduced or refined to a few valuable species or replaced by a simplified plantation forest; a grassland is converted into a grain field, and so on. By such means, man has managed to obtain high yields of species he favours. But these yields are likely to be either temporary or highly fluctuating, unless factors which, in a natural community, lead to stability and productivity are compensated for by artificial means.

These general observations on ecological relationships have a direct bearing on development. If their lessons are ignored, entirely unexpected consequences can often result from what are intended to be straightforwardly beneficial activities. The removal of a complex tropical rainforest and the substitution of a simple plantation of oil palms, for example, sets in motion a great number of ecological forces. The habitat for the enormous variety of insect species that occurred in the original forest will have been destroyed. Moreover, a habitat suited to those insects that feed upon oil palms will have been greatly expanded. Such insects can be expected to increase to pest proportions initially, and do damage to the palms. In the normal course of events, however, those insects that prey on the oil palm feeders, or the organisms that parasitize them, will also increase, but more slowly. Following the initial increase in pests, there will be a decline as their natural enemies catch up with them. If there is some reasonable degree of diversity in the environment surrounding the plantation, habitats may be available for a wide variety of enemies of the oil palm pests, and alternative species may be available for them to prey upon when the oil palm pests are reduced in numbers. Under these circumstances, an equilibrium may be established with both pests and predators occurring at levels that do no great damage to the plantation. However, if the environment has been greatly simplified and available habitats are widely reduced in biological variety, then it may not be possible for either pests or predators to stabilize. Fluctuations in the numbers of both will then occur, with years when pests are abundant and years when they have been reduced by their predators.

If, into this artifical situation a further element is introduced, the use of a broad-spectrum pesticide such as DDT, another course of events may follow. The insecticide, when first applied, will be destructive to pests and predators alike, and both will be reduced to low levels. The habitat will, however, remain highly suitable for the crop pests, and those forms that have escaped the insecticide, or have some natural resistance to its effects, will begin to build up in numbers. Freed initially from their predators, they will again increase to high levels. The predators will increase more slowly, in part because their viability is based on a stable abundance of prey and also because, for many species, predators are likely to be less fecund than plant eaters. Furthermore, the surviving predators will feed upon prey already containing DDT and tend to get a higher average dosage than the prey species. They are therefore more likely to be entirely eliminated by the insecticide. If this happens, as is likely when DDT is applied repeatedly, then the crop pests may be entirely freed from their natural enemies, and control of their numbers will become entirely dependent upon repeated applications of pesticide. If the pest species develops a resistance to DDT (as is more the rule than the exception), then after a few applications of the pesticide a new generation of pests which cannot be controlled by DDT will have been produced, a generation, moreover, freed from all natural predation.

Where a situation is allowed to develop to this point, new and more effective insecticides may be sought. These may have an even wider biological impact on other species and will probably also result in the eventual production of generations of pests resistant to the new insecticide. The ultimate consequences, apart from the harmful side-effects, may well be an extremely expensive, continuous control operation if crop production is to continue—or else a decision to return to more natural conditions and a restoration and enhancement of natural or introduced biological controls. This could involve the use of selective, rapidly degradable pesticides, or a manner of using pesticides which will affect only the pest species, allowing the predators and other enemies of the pest to maintain normal population levels. The manager, at this point, assumes a role of a selective, additional predator upon the pest insect, and a new equilibrium in which the pest populations are kept at low and tolerable levels should be obtained. It need hardly be pointed out, however, that this more rational approach to the problem of crop pests is not widely applied. Development efforts are still directed largely at the temporary relief provided by simplistic application of broad-spectrum, persistent pesticides. The topic is explored in greater detail in Chapter 6.

(8) BIOTIC SUCCESSION AND LAND REHABILITATION

The ability of an ecosystem to tolerate human use and to recuperate from human abuse varies greatly with climate and biological factors. The recovery of communities from fire illustrates this. In areas that usually have high rainfall and humidity, for example, naturally occurring fires are rare. In unusually dry years, however, large fires may burn. The immediate results can be disastrous because of the great volume of accumulated fuel in the vegetation. However, the results of the fire are seldom long lasting. A process known as biotic succession begins, and follows a more or less regular and predictable sequence. In forest areas a weed stage commonly takes over first on burned ground. After a few years, shrubs and trees of species that germinate on open, unshaded ground will move in and shade out the weeds. The early shrubs will be replaced by the light-tolerant or light-demanding trees. These in turn provide cover for those shade-tolerant species of trees that can seldom establish themselves in the open. As these grow, they will replace the shade-intolerant species. Eventually, and provided a seed supply is relatively near, the species characteristic of the original forest will occupy the ground; and its structure, physiognomy and botanical composition may remain stable for an indefinite period of time. This relatively permanent form of vegetation is known as *climax*. The vegetation that precedes it in occupying the open ground is *successional*. Succession can be diverted or delayed. Thus, if bare ground in the tropics is invaded by tall grasses of such species as *Imperata cylindrica,* they may hold the ground and resist invasion by trees and shrubs for some time. If the grasslands are burned frequently, invading trees and shrubs can be indefinitely excluded. A so-called 'disturbance' climax or 'fire' climax may be the result.

In desert regions, naturally occurring fires are virtually non-existent since there is inadequate vegetation and fuel to carry a fire. In sub-humid and semi-arid regions, fires are most frequent because both adequate fuel and dry seasons occur. Vegetation recovers from fire most quickly in areas with higher rainfall, deeper soils and a greater variety of species. In semi-arid regions, however, recovery may be slow since succession is slowed down by the absence of adequate soil moisture and the relatively small number of species adapted to arid conditions. In some circumstances, recovery of the original vegetation may not take place at all, as, for example, when the plant species involved are growing near the arid extremes of their natural range. Similarly, in cold regions the effects of fire are long lasting. Tundra fires in Alaska and Canada destroy the lichen vegetation which is a major part of the climax. Recovery of the lichens takes many decades if it occurs at all.

Other disturbing factors produce results similar to those brought about by fire. Grazing of climax grasslands that have not recently supported grazing animals results in a shift in the species composition of the vegetation towards those plants best able to withstand cropping. If grazing is removed the less tolerant plants will recover, but if grazing pressure is increased to high levels virtually all vegetation may be killed. However, recovery will usually take place along predictable successional pathways if the grazing animals are reduced in numbers or removed entirely. Similarly, land clearing or any other process that removes the original vegetation will set in motion successional processes which will eventually lead towards recovery of the climax.

The climax vegetation is that best suited to the climate, soils, physiography and biotic influences of a region. Since biotic succession leads towards this climax it is difficult, without constant management, to maintain an area in vegetation which is successional by nature or is exotic to the area concerned. The continuing battle against weeds on agricultural ground is a battle to arrest normal successional processes. The efforts to keep savanna grasslands free from woody vegetation represent a similar struggle against succession.

It should be added that the rates of succession will vary, of course, with soil fertility. If the latter is lost, there will be, at best, a very slow rate of establishment and succession of new vegetation.

(9) RESILIENCE OF POPULATIONS AND COMMUNITIES

The ability of a community to recover from disturbance is a measure of its resilience. Some ecosystems are highly resilient, others are exceedingly fragile. As noted above, communities of arid or cold regions may be fragile, and their ability to recover from disturbance is limited. Communities in more favourable environments tend to be resilient. There are, however, exceptions to this rule, such as the highly specialized communities which have developed on some isolated islands and high mountains.

Islands which are located at considerable distances from the mainland can be reached only by those species which possess considerable powers of dispersal: plants with light, wind-borne seeds or spores, micro-organisms with resistant eggs or spores, flying animals, species that can withstand ocean voyages on floating rafts of natural vegetation, plants with water-resistant seeds, etc. Most mainland species cannot reach distant islands, and in consequence, islands have developed unique, specialized biotic communities. Considerable internal evolution has taken place on islands which have been isolated over millions of years, such as the Hawaiian chain or the Galapagos. Entirely new species, not found on the mainland, have evolved. Island communities, therefore, are highly susceptible to disturbance. Where man has introduced mainland species of plants or animals, particularly very adaptable ones, these have often spread through the island communities, displacing the native species which evolved in the absence of such competition from their mainland relatives. Island animals may have developed no behavioural responses to predators, if predators have not evolved on or previously reached the islands. In some cases they may be entirely 'fearless' in behaviour, and some island birds have even lost the power of flight. Introduced predators—rats, cats, mongooses, etc. and man himself—can therefore easily cause their extinction. Island plants may have little resistance to grazing use and may be wiped out by introduced goats, sheep or cattle. Island vegetation may be invaded and native species displaced by vigorous species from the mainland, as has happened with *Lantana camera* in Hawaii. Attempts to develop island environments can have disastrous effects if they lead to decimation of native flora and fauna, which is not only an important scientific and recreational loss in itself, but can also lead to accelerated soil erosion, breakdown in soil fertility, and total disruption of island ecosystems. There are great numbers of once fertile islands which have been converted to virtual deserts by misplaced exotic introductions: some of the Grenadines in the West Indies are notorious examples.

High mountain tops represent environments which differ markedly from those of the lower slopes. Alpine areas in the tropics often have more in common with arctic regions than with the humid tropical lowlands. Where high mountains are isolated from one another, like Kilimanjaro, Mount Kenya and Mount Elgon in East Africa, their upper altitudes have the same degree of isolation as oceanic islands. Such areas can be reached only by the more mobile animals adapted to their climates and vegetation, and by plant species whose seeds can be dispersed by wind or carried by birds. Like islands they become centres for local evolution and, like those of islands, their plant and animal communities can be highly fragile, because of their isolation and the rigorous climatic conditions under which they live. An equivalent care must be exercised in their management if their productivity, usefulness and biological values are not to be destroyed.

(10) ECOLOGICAL BALANCES AND SURVIVAL THRESHOLDS

Any ecosystem, no matter how resilient, can be pushed to a 'point of no return' or, more exactly, to a threshold beyond which limiting factors become so severely operative that recovery, in periods meaningful in the human time-scale, becomes impossible.

For example, soil erosion on mountainsides may reach a point where bedrock is exposed. When this occurs only the slow processes of primary biotic succession, operating over centuries or millenia, can build back the soil and vegetation. Another 'threshold' process, typical of many tropical areas, is soil laterization, which can only be arrested econômically at certain early stages. Once a tough layer of laterite is exposed or formed, recovery may no longer be feasible. Communities can repair themselves, with moderate protection, up to a certain point. Beyond that level of abuse or disturbance biological repair becomes, in human terms, an intolerably slow process.

Consistent overcropping of any species can reduce its numbers to levels from which recovery is no longer possible. Killing the last individual animal or cutting the last tree of a species may not be needed to bring about extermination. If animal populations are reduced to a level at which effective breeding does not occur, or if the habitat of a species is destroyed, recovery will not take place, even though some individuals may for a time survive. Similarly for plants, if a minimum area to ensure regeneration of certain species is not maintained, they will eventually die out. The number of individuals or the size of the habitat needed to assure survival of species has not yet been generally determined. The threshold apparently varies with the genetic make-up of the species, its habitat requirements and behavioural characteristics. Herding or flocking animals or schooling fish may require some minimum 'critical mass' for breeding to occur or for the young to survive. This critical mass can be considerably larger than one would otherwise expect, due to the important behavioural characteristics of the species, or its relationship to predators.

(11) REFERENCES

For the purposes of this chapter it has seemed inappropriate to interrupt the flow of discussion by identifying in detail the individual sources from which some of the ideas and examples have been drawn. But the authors' debt, both here and elsewhere in this book, to the major contributions to the science of ecology listed below, is gratefully acknowledged. To all who seek a deeper understanding of the subject, these key studies, all but two of them published within the last 15 years, are most strongly commended.

Aubert de la Rue, Edgar, Bourlière, François, and Harroy, Jean-Paul (1957). *The Tropics.* Alfred A. Knopf, New York.

Borgstrom, Georg (1969). *Too Many: A Study of Earth's Biological Limitations.* Macmillan, New York.

Clements, Frederic E. (1963). *Plant Succession and Indicators.* Hafner, New York.

Dansereau, Pierre (1957). *Biogeography: an Ecological Perspective.* Ronald Press, New York.
Darling, F. Fraser, and Milton, John P. (1966). *Future Environments of North America.* Natural History Press, Doubleday and Co., New York.
Darling, F. Fraser, and Milton, John P. (1968). *Environmental Conservation.* John Wiley and Sons, New York.
Elton, Charles (1927). *Animal Ecology.* Sidgwick and Jackson, London.
Elton, Charles (1958). *The Ecology of Invasions.* John Wiley and Sons, New York.
Elton, Charles (1966). *The Pattern of Animal Communities.* John Wiley and Sons, New York.
Eyre, S. R. (1963). *Vegetation and Soils.* Edward Arnold, London.
Forrester, Jay W. (1971). *World Dynamics.* Wright-Allen, Cambridge.
Graham, Edward H. (1944). *Natural Principles of Land Use.* Oxford U. Press, New York.
Knight, Clifford B. (1965). *Basic Concepts of Ecology.* Macmillan, New York.
Kormondy, Edward J. (1969). *Concepts of Ecology.* Prentice-Hall, Englewood Cliffs, N.J.
Massachusetts Institute of Technology (1970). *Man's Impact on the Global Environment.* MIT Press, Cambridge.
Nicholson, E. M. (1970). *The Environmental Revolution.* Hodder and Stoughton, London.
Odum, Eugene P. (1959). *Fundamentals of Ecology.* 2nd ed. W. B. Saunders, Philadelphia.
Odum, Eugene P. (1969). *Ecology.* Holt, Rinehart and Winston, New York.
Reid, George K. (1961). *Ecology of Inland Waters and Estuaries.* Reinhold, New York.
Richards, P. W. (1957). *The Tropical Rain Forest.* Cambridge U. Press, London.
Thomas, William L., Jr., ed. (1956). *Man's Role in Changing the Face of the Earth.* Univ. of Chicago, Chicago.
Van Dyne, George M. (1969). *The Ecosystem Concept in Natural Resource Management.* Academic Press, New York.
Watt, Kenneth E. F. (1966). *Systems Analysis in Ecology.* Academic Press, New York.
Watt, Kenneth E. F. (1968). *Ecology and Resource Management.* McGraw-Hill, New York.

CHAPTER 3

Development of Humid Tropical Lands

(1) CHARACTERISTICS OF TROPICAL FOREST ECOSYSTEMS

The more humid regions of the tropics are characterized by the most complex and diverse forests on earth. These are adapted to prevailing conditions of high precipitation and humidity, and year-round warm temperatures. Within the tropical region, however, there are marked differences in climate, particularly those associated with changes in elevation, and also major differences in soil. All of these give rise to changes in vegetation. Tropical forests are, therefore, not uniform in species composition nor in structure and the general term 'tropical rainforest', as popularly used, covers many different types of forest.

Climate, soil and biota interact to make the tropics, particularly the rainforest regions, among the most difficult areas of the world for intensive human settlement and economic development. Nevertheless, where the soil fertility can be maintained under agricultural use, as in areas of South-East Asia, high population and cultural levels have been achieved for a long period of time. It is dangerous, however, to assume that the techniques that worked in one area of the tropics can necessarily be applied to another where soils, vegetation and human cultures are different. Even greater danger lies in assuming that the methods and techniques that work well in the temperate zone can be transferred without modification to the tropics. Both of these assumptions, in the past, have led to serious land misuse and to long-term economic and environmental loss.

Within the rainforest region of the tropics, which is taken here to include rainforest, wet forest and moist forest, as described by Holdridge (1967), precipitation is usually high, above 200 cm annually and in some places exceeding 800 cm. More importantly, precipitation exceeds (or equals) the potential evapotranspiration so that a shortage of soil moisture does not seasonally inhibit plant growth. Although there may be short dry periods when rain does not fall, there is no season in which the soil dries out and plant growth ceases. Plants in this region are highly susceptible to even short periods of insufficient moisture. Generally speaking, except at higher altitudes, there is no month of the year in which mean temperatures fall below 18 °C.

Within the dry forest region there may be one or more dry seasons during the year sufficient to inhibit plant growth. Annual rainfall, however, is usually above 100 cm and is usually equal to at least half the total annual potential evapotranspiration. There is thus a season of heavy rainfall and abundant plant growth, and great monthly variability in precipitation, which is also less reliable than in the humid forest regions. Temperature, also, tends to be more variable although the monthly mean does not usually fall below 18 °C.

In addition to these temperature and precipitation factors, tropical climates are marked by a high incidence of solar radiation and relatively little variation in length of day. Thus, even at high elevations in the tropics, where temperatures may be low and snow replaces rain, the conditions for plant growth differ markedly from those of temperate or arctic climatic regions.

Forests consequently prevail throughout the region and are absent only where soil conditions or lack of soil prevent tree growth or where it has been inhibited by disturbances, such as fire or land clearing. Tropical forests are noted for being the most diverse, luxuriant and productive, in terms of gross organic matter, on earth (Figure 4). They are therefore of extremely great scientific interest, a reservoir of genetic materials unequalled elsewhere.

Animal life is equally abundant and diverse in the tropics. In the more humid areas, the number of species of large and conspicuous mammals may be relatively small, but there are many more species of other vertebrates, particularly birds, than in comparable areas of the temperate zone. Invertebrate, and particularly insect species, are exceedingly numerous. Biotic factors more commonly limit population growth in the humid tropics than in other regions on earth. A high percentage of trees and other plants

Figure 4. Rain forest in Thailand showing a small segment of the great diversity in plant life that characterizes natural communities in the humid tropics. (Photograph by F. Vollmar: courtesy World Wildlife Fund).

depend upon animals both for pollination and for dissemination of their fruits and seeds. The degree of interdependence is strong: often a single species of insect, bird or bat will be the sole pollinator of a particular plant. Without the animal the plant cannot survive and, conversely, the animal may be entirely dependent upon the plant species. In the drier tropics the abundance of large wild mammals is one of the more spectacular features, especially in Africa and Asia and, to a lesser extent, in undisturbed areas of Latin America.

Climate and biota interact with the rocks and minerals of the earth's crust to form and to modify the soils of the tropics. Altogether they form a tightly woven ecosystem which presents a major obstacle to those seeking to carry out economic development. Thus, without the natural vegetation the productivity of the soil may be quickly lost; without the animal life, the vegetation cannot be perpetuated. It is difficult to modify the total forest ecosystem, particularly in humid regions, to make it more useful for commodity production, and if it is destroyed, the man-made ecosystems used to replace it are often highly unstable.

Through large parts of the humid forest region the soils are of the group now termed *ferralsols* (previously called latosols or lateritic soils).* These are heavily weathered soils from which the more soluble minerals have been leached. The remnant materials are those silicate clays resistant to leaching, along with iron, aluminium or manganese oxides. Their continued fertility is dependent upon the screening presence of forest vegetation, which modifies the leaching and weathering action of high temperatures and precipitation, and upon the ability of deep-rooted vegetation to restore minerals to the soil surface from the deeper layers of soil or parent material.

Most soil nutrients are tied up in the biota. Forest litter decomposes rapidly, the process being assisted by an abundant soil fauna, and its minerals are quickly brought back into recirculation. Plants are continually incorporating carbon compounds into the soil and micro-organisms do the same with the atmospheric nitrogen which they capture. However, but for these processes, the soils are relatively sterile, although the existence of a

Figure 5. Severe erosion following forest clearing, subsistence farming and grazing in the Semien Mountains of Ethiopia. (Photograph by F. Vollmar: courtesy World Wildlife Fund).

* The soil terminology employed here is that used on the world soil maps under preparation by FAO, UNESCO and the IUSS, and is described fully in Dudal (1968).

luxuriant forest gives a misleading impression of soil fertility. The fertility is in fact tied up with the forest system, disappearing if the forest is replaced by more widely-spaced, shallow-rooted and short-cycle crops, which expose the soil to weathering (Figure 5).

Attempts to clear this kind of soil for agriculture have consequently been marked with little success. If the vegetation is cut down and burned or its litter is allowed to decompose *in situ,* the nutrients that were present in the biota will be restored to the soil. This can sometimes give a high immmediate fertility so that, in the first year or two after clearing, crops can be raised successfully. However, if the soils continue to be exposed to the sun and rain, bacterial action quickly breaks down the organic material and leaching removes the nutrients from the surface soil. Soils often become compacted and lose air space and water holding capacity. Thereafter, fertility virtually vanishes, and crops will fail. Certain of these soils, the true laterities, will bake into hard, almost indestructible crusts of iron and aluminium oxides upon exposure to sunlight and rain. In others, a lateritic crust exists in the subsoil and is exposed when the soils are subjected to surface erosion.

Where these lateritic crusts are concentrated and exposed due to natural or induced erosion, they destroy agricultural values and may form vast sterile stretches, usually on upland plateau landscapes. In some cases the indurated or ironstone-rich horizon is overlain by topsoil and can be used for some perennial crops such as coffee, but cacao, plantains and clean-cultivated annual crops are liable to produce erosional losses, with eventual disappearance of what useful soil may have existed. Eroded soils too shallow for crops may be suitable for grass. Thus, while such soils are not worthless, they require careful management and their productivity is limited (Aubert, 1963).

The second most widely distributed group of soils in the humid tropics are the *acrisols* (formerly known as red-yellow podzolic soils of low base status, *terra roxa estruturada,* etc.). These, unlike the ferralsols, show a more or less distinct horizon development, with a light-textured surface soil and a heavier clay-rich subsoil. Although not as strongly weathered as ferralsols, the rate at which weathering adds new nutrients is exceeded by the rate at which these minerals are leached away. The clays, often kaolin, have a low base-exchange capacity. Consequently, as with the ferralsols, the continued fertility of these soils depends on the activity and shielding effect of the forest cover. When this is removed fertility is quickly destroyed.

Not all soils developed in the humid tropics are infertile. Noteworthy exceptions of high fertility and stability are certain *andosols,* developed on some types of volcanic ash beds, the *paramo* soils of the subalpine region, *nitosols,* developed from basic volcanic rock, and local areas of *grumosols* and *rendzinas*, developed from calcareous rocks and limestones. Although man has occupied most of these areas of high productivity, the intensity of land use and of cultivation usually does not take full advantage of their productive capacity.

Alluvial soils in the humid tropics vary enormously in fertility depending largely upon the source of their parent material. Those derived from the erosion of volcanic highlands or areas of geologically young mountain ranges, for example in Java, Indonesia or the Guayas River region of Ecuador, may develop high nutrient status upon weathering; those from ancient shield rocks and from soil areas dominated by ferralsols and acrisols are likely to be little superior to the upland soil areas from which they originated. Thus, the alluvial soils of much of South-East Asia, derived from volcanic or young uplands, tend to be highly fertile, whereas many of the alluvial soils of central Africa are quite the reverse. In Brazil the 'black' rivers, such as the Tapajoz, carry little suspended material, and alluvial soils are scarce. In the 500,000 square miles of the Amazon Basin, only 25,000 square miles of alluvial flood-plain soils (varzeas) are found (Sternberg, 1964), and these are seasonally flooded. The limitations and potentials of tropical alluvial soils have been reviewed by Edelman and Van der Voorde (1964).

The danger of soil erosion in the tropics is considerably aggravated by the circumstance that temperature conditions make cultivation of economic plants possible in mountainous regions. In fact, regions of steep topography may have the most fertile soils if enriched by volcanic outpourings or if developed over igneous rocks rich in mineral nutrients. Hence tropical uplands with slopes as steep as 40 degrees are widely farmed and harbour dense populations in Latin America, Africa and Asia, both historically and at the present. Soil erosion poses the greatest problem in cultivated zones where the rainfall pattern is seasonal, where monthly rainfall is about equal to or less than evapotranspiration potentials, and where rain falls as intense showers. The cultivation of short-cycle crops such as cereals and tubers on such soils is accompanied by very rapid erosion. Soil losses are particularly catastrophic if the first rains falling on the ploughed and seeded slopes are intense downpours. The spread of pastoralism into such areas of steep topography brings further danger of erosion, since overgrazed and trampled slopes are particularly vulnerable to soil loss.

Areas which suffer these types of soil erosion are found in the subhumid, steep terrain of southern Mexico and central America, traditionally dominated by subsistence farming based on corn and beans; in the potato and cereal growing areas of the Andes; and in the upland cereal growing regions of Asia Minor and the southern Mediterranean countries. Drier zones in mountainous terrain, which are only occasionally cultivated because of the unreliability of rain, are also liable to serious erosion, especially where land use is dominated by charcoal gathering and goat herding—activities which permanently deplete the scant vegetative cover. This has led to landslides and the total denudation of mountainsides in Peru (Tosi, 1960) and several other countries. Such destruction of productive land by erosion certainly constitutes one of the most serious resource management problems in the world. Corrective action involves all aspects of land use and needs to be carefully planned and administered: this particularly applies to terracing, which is labour intensive both in construction and management.

In tropical and subtropical Australia, the relationship between natural vegetation and the type of soils has been carefully studied. Webb (1968) has shown how closely the geology and soil mineral content are linked with the type of rainforest or other forest vegetation. In Guyana, Richards (1961) noted the dominance of mixed tropical rainforest on the prevalent, permeable ferralsols. On more localized soil types, single species tended to be dominant: *Mora excelsa* on alluvial soils, Morabukea forest on heavy-textured dark-red clays, Greenheart forest on light brown sands, Wallaba forest on tropical podzols derived from quartz sand. Others have recorded distinctive vegetation types confined to lateritic crust areas, rendzinas, etc. Further correlation will no doubt permit more rapid identification of soil potential from studies of plant associations.

In the sub-humid, dry forest region the trees are characteristically 'rain-green', shedding their leaves during the dry season. Evergreen species also occur, but deciduous forms predominate. The vegetation is not so complex as in rainforest, epiphytes are less common, and the number of species of plants per unit of area is reduced. Dominance by a single species of tree becomes more frequent over some areas, and in many areas a few species share dominance. Because dry forest is subject to fire in the dry season, it has been replaced by savanna in many areas, and fire-tolerant species of trees are likely to survive best. Soils vary from the *ferralsols* and *acrisols* to more fertile *luvisols* (red-yellow podzolic soils of high base status) and, in some areas, to *castanozems* (chestnut soils), or to *phaeozems* (prairie soils). Red and yellow acid quartz sands (quartz *regosols*) are often widespread.

A widespread vegetation in the dry forest region of South America is the *cerrado,* dominated by low, contorted trees and shrubs. Although this occurs in a rainfall regime capable of supporting high forest, the persistence of fires, and a lack of a proper balance of plant nutrients in the soil is believed to retard forest growth. Where richer luvisols or rendzinas occur in this region, and fires are not frequent, forests grow to normal height and density. The dominant soils, however, are ferralsols and quartz regosols. Similarly, in the rainforest region of Brazil, quartz regosols are frequently occupied by an open, apparently drought-resistant evergreen sclerophyll forest, the *caatinga,* which represents a response to deficiencies in soil nutrients, since the soil is not deficient in water. By contrast, the deciduous type of *caatinga,* an open raingreen forest, occurs in areas where the dry season is severe and evapotranspiration greatly exceeds precipitation, but where the soils offer a more balanced array of nutritional elements.

(2) BIOTIC SUCCESSION

Within the humid tropics, vegetation is quick to colonize bare ground if the soil remains fertile. Vegetation growth is rapid and net productivity or total gain in plant biomass may be very high where soils are not greatly disturbed. Where disturbance is great, vegetation may still move in and occupy the ground in a relatively short period of time, and succession may proceed quite rapidly. Thus, following the volcanic eruption of Krakatoa, resulting in total destruction of life on the island, plants moved in from surrounding islands within a few years and, within 25 years, woodland or savanna occupied much of the surface. The substrate on Krakatoa was of high potential fertility, but where fertility is destroyed through the leaching and baking of soils, invasion may be retarded and in extreme cases, such as in highly laterized areas, only specialized, tolerant plants may be able to colonize the ground.

The ability of tropical vegetation to quickly reclaim cleared ground has been the bane of agricultural development and, at the same time, the secret of the continued success of shifting cultivation, involving a long fallow period after a brief interval of cultivation. Where the farmer or pastoralist attempts to keep ground open and free from invading plants a continual and ever more difficult battle must be waged. Except where intensive agriculture is socially and economically feasible, it is usually a losing battle.

The tendency of secondary forest of various kinds to hold its ground over long periods of time has been widely observed. It has been noted, for example, that after 5 or 6 centuries the vegetation of cleared areas around Angkor in Cambodia still does not fully resemble the surrounding areas of undisturbed climax.

Although vegetation invades cleared areas, and succession is normally rapid, recovery of primary forest following major disturbance is often slow. In part, the extreme complexity of the climax rainforest and the intricate network of plant–animal relationships slows down complete recovery from disturbance. The loss of a species of bird can prevent dissemination of seeds of a particular tree, and the loss of an insect species can prevent its pollination.

Although much remains to be learned about succession in humid tropical forests, considerable information is already available. Budowski (1965) has summarized information for Central America as in Table 1 below. The tendency to develop from simple pioneer communities towards the diverse flora of the climax forest is well shown. Budowski has also noted that many trees of high value for lumber, such as mahogany (*Swietenia*), *Cedrela* and *Ceiba,* which may appear as dominants in otherwise climax forests, are actually long-lived species from late secondary succession and may be unable to regenerate under climax conditions.

Although succession would normally proceed from herbaceous plants to low woody plants, to low trees, and on to taller woodland and forest, it may take other routes. Vigorously invading grasses, such as the lalang or alang-alang (*Imperata cylindrica*) of South-East Asia, may enter forest clearings, particularly where soil fertility has been depleted by agricultural use, and if periodically burned will hold the ground for long periods. Where dry seasons occur, such grasses burn readily, may remain dominant almost indefinitely and further invade the burned forest edges. Since they are not only of low value for pasture, but also almost impossible to eradicate, their invasion has marked the end of various pastoral schemes.

(3) DIVERSITY, STABILITY AND RESILIENCE

The diversity of tropical forests, already emphasized, is well-illustrated by one small volcano in the Philippines (Mount Makiling), on which more species of woody plants have been recorded than in the entire United States. This diversity, in combination with generally favourable growing conditions, allows vegetation, *per se,* to quickly colonize and invade any bare and fertile ground, and forest to reform itself rapidly after disturbance. At the same time, complete recovery of climax forest following major disturbance has been observed to occur only very slowly.

Diversity has other consequences. If great numbers of any single species do not occur in an area, its predators and parasites will likewise not be abundant or concentrated. Each plant-eating animal may have a great variety of species that feed upon it, but each of these in turn will have a variety of alternate prey species on which it may feed when the first becomes scarce, and each will also have its array of enemies.

Compared with other ecosystems, those of the humid tropics, at least in the case of mature biotic communities, are buffered against major fluctuations in the relative abundance of any one species. They show a high degree of stability from year to year, changing continually with the germination or birth, growth or death of individuals and yet maintaining the same balance of species and overall structure and composition.

Table 1. Characteristics of arboreal components of seral stages in tropical American humid forests (after Budowski, 1965).

	Pioneer	Early Secondary	Late Secondary	Climax
Age of communities observed, years	1–3	5–15	20–50	More than 100
Height, metres	5–8	12–20	20–30, some reaching 50	30–45, some up to 60
Number of woody species	Few, 1–5	Few, 1–10	30–60	Up to 100 or a little more
Floristic composition of dominants	Euphorbiaceae, *Cecropia*, *Ochroma*, *Trema*	*Ochroma*, *Cepropia*, *Trema*, *Heliocarpus* most frequent	Mixture, many Meliaceae Bombacaceae, Tiliaceae	Mixture, except on edaphic association
Natural distribution of dominants	Very wide	Very wide	Wide includes drier regions	Usually restricted, endemics frequent
Number of strata	1, very dense	2, well differentiated	3, increasingly difficult to discern with age	4–5, difficult to discern
Upper canopy	Homogeneous, dense	Verticillate branching, thin horizontal crowns	Heterogeneous, includes very wide crowns	Many variable shapes of crowns
Lower stratum	Dense, tangled	Dense, large herbaceous species frequent	Relatively scarce, includes tolerant species	Scarce, with tolerant species
Growth	Very fast	Very fast	Dominants fast, others slow	Slow or very slow
Life span, dominants	Very short, less than 10 years	Short, 10–25 years	Usually 40–100 years, some more	Very long, 100–1000, some probably more

Table 1—*continued*

	Pioneer	Early Secondary	Late Secondary	Climax
Tolerance to shade dominants	Very intolerant	Very intolerant	Tolerant in juvenile stage, later intolerant	Tolerant, except in adult stage
Regeneration of dominants	Very scarce	Practically absent	Absent or abundant with large mortality in early years	Fairly abundant
Dissemination of seeds of dominants	Birds, bats, wind	Wind, birds, bats	Wind principally	Gravity, mammals, rodents, birds
Wood and stem, dominants	Very light, small diameters	Very light, diameters below 60 cm	Light to medium hard, some very large stems	Hard and heavy, includes large stems
Size of seed, or fruits dispersed	Small	Small	Small to medium	Large
Viability of seeds	Long, latent in soil	Long, latent in soil	Short to medium	Short
Leaves of dominants	Evergreen	Evergreen	Many deciduous	Evergreen
Epiphytes	Absent	Few	Many in number, but few species	Many species and life forms
Vines	Abundant, herbaceous, but few species	Abundant, herbaceous but few species	Abundant, but few of them large	Abundant, includes very large woody species
Shrubs	Many, but few species	Relatively abundant but few species	Few	Few in number but many species
Grasses	Abundant	Abundant or scarce	Scarce	Scarce

Simplification of tropical ecosystems, such as necessarily accompanies agriculture or pastoralism as well as intensive forest culture, must always tend to upset the balances developed in the mature biotic community. Where only one or a few species of plants are encouraged to grow in place of the diversity of the natural forest, an environment is created that is highly favourable to the rapid increase of their predators or parasites.

The consequences have already been briefly discussed in the previous chapter (pp. 43–45) and will be examined in greater detail in Chapter 6, in connection with the ecological impact of agricultural production technology.

(4) LIMITATIONS FOR DEVELOPMENT

The ecological factors described above all exercise limitations on economic development of the tropical forest region. In various ways they account for the depressing series of failures that have too often accompanied development efforts in the past (Figure 6). These efforts have typically concerned four activities—farming, forestry, road building and public health measures—and some of the conclusions to be drawn and ecological principles involved in each of these fields of interest can usefully

Figure 6. Extensive clearing of tropical forests such as this area in Malaysia may result in the loss of long-term values for transitory gains. (Photograph by Oliver Milton: courtesy World Wildlife Fund).

be considered here, although more detailed discussion of many of the problems follows in later chapters.

(a) Agricultural and Pastoral Development

Traditional tropical agriculture has a long record of success. It has adapted to the soils, vegetation, and climate of the tropics over centuries of trial and error. Two major patterns are evident: those systems based on restricted areas of highly fertile soils, and those based on the more general run of *ferralsols, acrisols* and other 'problem' soils of the tropics. Examples of the first pattern are the long-established use of *fluvisols* (notably river alluvium) in the Asian tropics, and the use of terrace-based agriculture to cultivate the highly productive *andosols* of the same region. Both take advantage of a fertile soil which is periodically enriched by the drift of river-borne or wind-borne materials. Both involve intensive care of lands which in turn produce high yields. Examples of the second pattern are in the slash-and-burn (*milpa, ladang, chitemene*) system of shifting cultivation. This takes advantage of the great supply of nutrients available in tropical vegetation, and in the resilience of tropical biota. It succeeds best where forest clearing is only partial and soils are not exposed to baking and leaching, where plantings are of perennial crops that give soil cover, and where the rotation is a long one, with a prolonged period under forest cover followed by a short period of crop production. It breaks down when soils are laid bare too frequently, or when the forest fallow period is shortened. This will occur when populations increase beyond the capacity of the agricultural system to provide support, or when a desire to engage in commercial marketing of products brings too great a pressure upon the productive capacity of the system.

A second example of a traditional system adapted to the poorer tropical soils is the village garden method of agriculture. This involves intensive care of a small plot. Large quantities of plant and animal wastes are added to the soil along with ash from wood fires. A variety of diversified plantings provide soil cover and round-the-year yields. Such a system is labour-intensive and needs a high degree of personal attention. As such, it is best suited to high value crops.

Successful modern approaches to agricultural development in the humid tropics are those that use the principles revealed in the past. In the words of Wright and Bennema (1965):

'The main problem facing the single farmer entering the humid tropics is how to adapt the natural fertility regime of the soil system to his own ends. If he removes suddenly the standing forest, burns the trash and clears the stumps, and bares the soil surface to the hot tropical sun while awaiting physical conditions suitable to him before sowing his first crop, the first harvest may be satisfactory, but subsequent harvests may decrease. This attack is abrupt, and the nature of the soil becomes changed in the space of a few years. In most soils the major change is the loss of plant nutrients; in some there may be important structural changes, while in others there is a change in aeration, water-holding capacity or other vital aspects. The fine balance of the component parts of the soil process has been thrown into disarray. By contrast, the farmer who approaches the virgin forest with a gentle hand, attempting to do no more than thin the canopy to introduce a tree crop of his choice in place of the original forest trees, finds little difficulty in establishing (by slow degrees) a farm that will endure. The soils remain virtually unchanged, and gradually a new organic regime supplants the old one, with a slower loss of fertility.'

Throughout a large part of Latin America, and to a lesser degree Africa, extensive tracts of humid forest are being cleared to make room for grasses for cattle (Figure 7). This is being encouraged, principally, by better communications and better technical means of removing large trees through mechanical devices that prove to be much more effective and economically advantageous than those based on the axe and saw. Many arguments are advanced in favour of this kind of development, and in several Latin American countries it has become the avowed policy that forest lands can advantageously be opened up for the establishment of pastures.

Figure 7. Extensive forest areas are cleared in Latin America to develop livestock pastures. Such pastures have short life spans. Ecuador. (Photograph by H. Jungius: courtesy World Wildlife Fund).

The results, at this stage, have been subjected to very little scientific investigation because political and social considerations, many of them stemming from a rapid growth of population, coupled with agrarian reform plans, do not favour an objective evaluation. However, as has been explained earlier, it is quite clear that in most of the tropics the good land has already been opened for agriculture and grazing, and what now remains is largely marginal land. The word 'marginal' is of course subjective and depends on a series of factors: however, it very often involves one or both of the following ecological conditions: (a) a soil which would not allow the maintenance of permanent pasture for more than a few years; and (b) a slope so steep that any establishment of grass followed by trampling of animals would bring about serious deterioration, particularly erosion, and again make it impossible to maintain permanent productive pastures.

The first factor tends to be far more serious because it is not easy to recognize. Opening an area of high rainfall by removing the forest and allowing the establishment of grasses gives apparently good initial results. The grass grows very fast and often displays a beautiful and homogeneous cover. The principal early problem, weed control, does not seem too difficult since it pays initially to have workers cut invading weeds with machetes. However, in succeeding years it usually proves impossible to fight the ever more aggressive encroachment of weeds, both annuals and more particularly perennials, due to a combination of factors of which the most important are the soil compaction, the diminution of organic matter and, linked with it, the weakening of the original grasses. The costs of keeping the pastures clear increase and it gradually becomes an uneconomic operation. After a time the pasture has to be abandoned and a new piece of forest is cleared. The occupation of the abandoned area by new secondary forest, of very little value, produces a landscape notably lacking in diversity. The original forest which had a high value for its genetic, scientific, educational and possibly also direct economic potential, may have been lost for ever.

There are exceptions to the above sequence of events and they correspond, particularly, to soils of exceptional structure and fertility which can take this kind of change in their cover without deteriorating. These are mostly certain alluvial soils as well as fresh volcanic soils of particularly good structure and composition. Such soils are, of course, also used in many countries for cultivation, especially of tree crops, such as bananas, oil palms and rubber. Nevertheless, it must constantly be borne in mind that pastures in the lowland humid tropics, except under specialized soil conditions, will not maintain themselves without a continued and costly input of management effort. Forest invasion of pastureland will occur naturally and proceed more rapidly whenever pastures are overgrazed. Fire is a tool that is useful for maintaining the grassland in the dry tropics if used carefully, but as will be noted in the next chapter, which discusses range and pasture management in more arid areas, the use of fire to maintain grasslands is not a panacea; it is rarely of value in the wet tropics. Great skill in the use of fire is required, and fire can no longer be used effectively if pastures are overgrazed. In consequence, the widespread conversion of forest to pasture in the humid tropics has been too often the first step, or an early step, in the destruction and eventual abandonment of the land resource. Commonly, also, conversion to pasture is attempted when crop farming has failed and can no longer be relied on to produce useful yields. It is then often only a matter of time until pastoral use also fails and what was once forest now becomes scrub. The sacrifice of high-value forests and other wildland resources for temporary gains in crops and pastures can seldom be economically justified. Except on limited areas of highly suitable soils, the higher the annual rainfall, especially if it exceeds 3,000 mm, the less likely is the possibility of establishing permanent pasture.

Pasture management in humid tropical lands also involves problems of animal nutrition. Grasses of high nutritional value may grow in recently cleared forest areas, where the soil has been enriched by the ash or decay of the original forest. However, this nutritional value will not be maintained. Most forest soils are not adapted to the production of high-protein quality forage, even though they may produce enormous quantities of cellulose, starch and lignin. Maintenance of pasture quality requires the addition of balanced fertilizers to the pasturelands to compensate for the poor nutrient balance in the soil.

There are, however, many problems associated with the application of chemical fertilizers to tropical soils; those encountered in Africa have recently been described by Phillips (1971). One of the more serious is the locking-up of phosphorus in unavailable form in ferralsols. The general lack of knowledge on fertilizer response under the great variety of soil-water regimes in the tropics is mentioned by Mukerjee (1963). Intercropping and mixed cropping may complicate fertilizer use, if a fertilizer needed by one species is toxic or harmful to the growth of an associated plant.

The complexities of tropical soils were analysed by Kellog (1949) some years ago, but his conclusions are as pertinent today as then. He emphasized the need to maintain shade over the soil surface (i.e. prevent high temperatures), the avoidance of unnecessary tillage, the use of mixed crops and rotations, the use of mulches and legumes, and the maintenance of organic matter. Soil management practices suitable for the better soils and climatic conditions in the tropics and subtropics are well enough developed to overcome any serious obstacle to agricultural development and higher production. The techniques of soil and crop husbandry for plantation agriculture, which is eminently suited to the more humid tropics, are quite advanced and many tropical soils respond well to chemical fertilizer and other management techniques. However, the economic infeasibility of increasing soil productivity through chemical fertilization and better management is still a major obstacle for the large majority of farm units in developing countries, namely the small family farms that have never employed these new inputs.

(b) Forestry

Tropical forestry, particularly in the humid regions, has lagged in development behind its temperate zone counterpart and, in consequence, forest utilization in the tropics has traditionally been a plundering operation with little concern for future yields.

Unfortunately, this tradition has persisted till today over wide areas. Too often logging precedes settlement, either as part of a deliberate plan, or through the inability of the government to control the movement of its peoples. The logging road, built for the purpose of removing a few high-value hardwood timbers, provides a means of access to a previously undisturbed forest. The timber cutters are frequently followed by the charcoal producers, who further the process of forest destruction, and they in turn by peasants or cattle raisers who remove the last traces of forest for the sake of what may well be only a temporary yield of crops or pasture.

Until recently there was little demand for the varieties of timber that tropical forests produce. The very diversity of the forests worked against the preservation of all these varieties, since the incentive for sustained yield production was lacking when commercially valuable trees were few and widely scattered, and the means for assuring their regeneration were generally unknown.

This situation has changed. It is now technically possible to make use of most species to meet different requirements for forest products. Furthermore, the demand for these products has not declined, but appears to be growing steadily, and may grow even more rapidly as developing nations move towards higher standards of living and higher levels of personal demand. In these circumstances the widespread elimination of tropical forests to make way for other forms of land use is short-sighted in both economic and ecological terms.

However, before an ideal situation is reached where sustained harvesting of mixed tropical forest can be achieved without devaluing or exhausting the entire forest capital, many obstacles must be removed; prospects at this stage look rather dim with the exception of those tropical forests, such as the mixed Dipterocarp forests of Malaya, where initially a high percentage of the tree species are valuable for timber and the marketing and processing situation is favourable. Elsewhere, sustained harvesting is commonly inhibited by very high transportation costs, the substantial investment that a large plant to process many trees implies, the difficulty of competing successfully in what is often a limited national market, the poor form of most of the tropical trees, the often seasonal nature of logging operations and the far from easy problems of storage. The only reasonable hope seems to lie in large industrial complexes that have found a ready market through diversification of their production to include timber, veneers, plywood, composition or chipboards and perhaps pulp. The refining of heterogeneous forests so as to limit the number of species to a few of high market value has not lived up to early expectations, witness research carried out in Trinidad and some African countries over the last twenty years. The costs of weeding and refining forests that do not have many species of commercial value has usually proved to be too high.

There is, however, another serious question that must be faced in most tropical countries. What is the relative value of the natural forest as compared to the plantation forest for the production of wood and other plant products? There seems little doubt that sheer volume of industrial wood production can best be produced on areas deliberately planted to one, or a few, well-adapted tree species—at least in the short run. H. C. Dawkins (1964) has reviewed the potential production from tropical forests, taking the maximum observed yield of the best species from the best sites. His conclusion was that manipulation of moist tropical forest of low to medium altitude leading to the most productive assemblage of species, would produce wood yields varying from 6 to 15 tons per hectare per annum, depending on the site. Replacement of the tropical forest by plantations of the best available local or exotic species, other than conifers or eucalypts, would produce yields of from 10 to 20 tons per hectare per year. However, plantations of conifers or eucalypts on the soils best adapted to them could produce yields of from 20 to 35 tons per hectare per year.

The intensively managed plantation of one or a few species, however, is subject to all of the ills that any form of monoculture will face in the tropics. Susceptibility to an increasing number and variety of diseases, parasites, insects or other animal pests is predictable; the potential for long-range decline in the productivity of the site caused by depletion of soil resources or impairment of soil structure is ever present. The costs of pest control and careful soil management, including fertilization, are at present usually justified on the grounds of the high value of the yields. But the full ecological costs of some of the existing pest control techniques have not yet been properly reflected in the cost-benefit equation.

The intensively managed plantation, like any other form of specialized agriculture, is virtually an exclusive use of the land. Most other values are necessarily sacrificed. Wildlife cannot be tolerated to the extent that it will damage the growing trees; recreational use is generally discouraged, and in any case the value of such forests for recreation is minimized by their structure and appearance.

Nevertheless, the spread of plantation forests and their continued success is in part a reflection of failure to understand and manage natural forests for their full range of values. Foresters and others concerned with wild land management have thus far been baffled in their efforts to make full use of the multiple benefits obtainable from complex and diversified forest ecosystems.

Some answers to the problem may be derived from the experience of Procter (1968) in Tanzania, where humid tropical forests occur mainly in mountain areas at fairly high elevations. Most of these have a recognized value in watershed management, maintaining the stability and quality of freshwater yields to streams and rivers. They also have value as production forests, but this is realized in the yield of various high quality natural timber species that cannot be produced in plantations. The maintenance of these natural forests depends upon forest wildlife, and the report by Procter fully supports the statement, made earlier in this chapter, to the effect that birds, mammals, insects and other animal life of the forest are essential for the continued reproduction of the trees. An excess of larger mammals may be damaging to regeneration, but in reasonable numbers they serve to keep the forest open and maintain sites suitable for the growth of some species. As natural forests, therefore, these areas have a realized value for watershed protection, recreation, wildlife and sustained yield of certain timber products. On the other hand, for the production of the great bulk of its industrial wood requirements, Tanzania relies and will continue to rely on forest plantations, usually sited close to population centres and to markets for their products. Similarly in Madagascar, tree plantations at present remove much of the pressure on the few natural forests which remain and which have quite exceptional value as natural reserves.

The fact is that forest plantations need occupy only a small percentage of the total forest land, and can be intensively managed for wood production with little conflict with other potential land values. On the contrary, they can contribute directly to reducing the pressure upon natural forests to supply all industrial needs, which is only possible through sheer volume of production if natural forest is subjected to wholesale exploitation. Plantation forests are most likely to be successful in fulfilling this useful role, in the tropics, if sited at medium to higher elevations in places where there is a high local demand for their products.

Unless strong measures are taken, it is likely that the next decades will witness the progressive destruction of the lowland tropical forest as a result of increased demand for land for agriculture and grazing. Since very little of this ex-forest land is capable of supporting permanent agriculture or grazing—or even a workable system of shifting agriculture or rotational grazing—it seems unavoidable that we will be faced with very large extensions of secondary forest. These will naturally vary in type, but they all tend to have certain characteristics in common: they are in a rapid state of development, dominant species are usually quick-growing until a certain age, and the timber is generally of a light and perishable quality. Moreover, the number of species, in comparison with the undisturbed forest, is much reduced and therefore secondary forests are much more uniform. Their management seems likely to be a main concern of tropical lowland forestry in the future. Industries will have to adapt themselves to their exploitation, in place of the traditional 'high-grading' schemes hitherto applied to mixed primary forests.

(c) Roads

Roads in the tropics are too often planned simply as a means for getting from one place to another at maximum speed and minimum expense. Little thought has been given to their overall effects upon regions through which they pass.

An obvious result of road construction through wild and unsettled areas is that it provides easy access to land not previously available to the settler. Legally or illegally, such land tends to be occupied quickly and used for whatever purpose the new occupant has in mind (Figure 8). Often such uses are completely disruptive to soils, vegetation and the entire ecosystem of the region suddenly made available. They seldom conform with the highest productive potentials of the land.

Initially, roads provide access for those who would hunt or trap wild animals or collect wild plants of economic value such as orchids. Areas within easy reach of roads will soon be denuded of the commercially useful and more easily captured or collected species, unless an adequate law-enforcement force is available for patrol. The next stage usually involves settlement by those who would establish various roadside enterprises, or who seek to clear and farm the land. Access is also provided to small timber operators who wish to harvest species of high commercial value. In the process of random logging and settlement, forests of high value for sustained yield production may be lost, and soils that are unsuitable for continuing agriculture or pastoralism are exploited, with consequent erosion or soil exhaustion.

The essential prerequisites for road building without environmental damage are that construction of new roads into hitherto inaccessible areas must be accompanied or preceded by land capability surveys, land use planning and effective land use control—including law enforcement and the provision of adequate extension services. Where the necessary surveys and land use plans cannot be undertaken, the responsible authorities must take

Figure 8. Road building in tropical forests such as in this area of Kenya may have effects extending far beyond the cleared area, particularly where land settlement cannot be carefully controlled. (Photograph by Kenya Information Service).

all necessary measures to control random exploitation of resources or unauthorized settlement. If none of the conditions mentioned can be met, it is ecologically irresponsible for any development agency to encourage the opening-up of new land through the construction of road systems.

(d) Health Problems, Pesticides and Herbicides

Settlement of the tropics was, during most of human history, hindered by the prevalence of disease. As a result, the tropical climate itself was considered unhealthy for man. This myth was dispelled when adequate sanitation and medical care could be made available. The dangers of tropical diseases have been further minimized by the discovery of a wide range of vaccines, prophylactics and antibiotics. The use of various pesticides against the insect vectors of disease has removed some of the more persistent limitations to human health, survival and population increase. However, few tropical diseases have been wholly eradicated, while much of the immunity and resistance of tropical peoples has now been reduced in contrast to the growing resistance of disease vectors. Dangers of major recurrences of disease await any breakdown in medical or sanitary systems.

The urgency attached to the alleviation of human suffering and the prolongation of human life has given a high priority to the immediate and widespread application of any techniques that show promise for the control of disease. Nevertheless, the environmental consequences of some of these techniques now need to be carefully considered. These consequences are discussed in detail in Chapters 6 and 7.

In the present context, it is sufficient to emphasize that it is in no way suggested that human health should be neglected or the use of pesticides and antibiotics (in effect, the broad-spectrum biocides) abandoned, but rather that it could now be stated, as a principle, that:

The use of pesticides and antibiotics should be carefully controlled during any economic development programme, in order to minimize undesirable side-effects; efforts to find effective substitutes for chemicals that have adverse effects on the environment should be pursued with the utmost vigour.

The use of herbicides (notably 2,4-D and 2,4,5-T) has unfortunately already had a major testing during the Indochina war. The consequences are potentially severe, and will be discussed in Chapter 6. Meanwhile, the only ecological guideline that can be suggested is one of caution in the use of any chemical that has potentially far-reaching environmental effects. It is also important to remember that the effects of any biocidal chemical in tropical environments are likely to be of an order of magnitude greater than those observed in the temperate zones.

(5) ALTERNATIVE SCHEMES FOR DEVELOPMENT

Most major schemes for economic development in the tropics have revolved around the opening-up of new lands and the exploitation of previously untapped resources. Often this is done with little or no planning for the best and maximum use of the land and resources to be developed. Too often, also, development alternatives are not given sufficient attention.

Virtually every tropical country has areas of land that have long been settled and maintained in one or another form of economic production. In most countries these lands are at present to some degree underutilized or inefficiently utilized. It is a recognized principle, as stated in Chapter 1 (p. 26), that concentration of development efforts in such areas is likely to produce much higher sustained economic benefit than the attempt to bring marginal areas into use. There are obvious exceptions to this principle, where previously unused lands of high capability can be opened up by the use of new techniques. However, far too much development money is being expended in activities that lead eventually to the casual and random destruction of lands and resources that it would have been better to have reserved for some more valuable long-term use.

Countries of the humid tropics are outstanding in the range of plant and animal communities they possess. This biotic diversity is of inestimable value to mankind for scientific, educational, and aesthetic purposes, a value which is bound to be enhanced with the passage of time, as such resources become more restricted.

Nations fortunate enough to possess undisturbed, or relatively undisturbed, natural areas therefore have a priceless asset. To destroy this through lack of foresight and lack of control over development processes would be irremediable folly.

All mankind, for all time, can be impoverished by the thoughtless action of a moment. Extinct species cannot be brought back to life. The most skilful manager of the future cannot restore the living communities of today if the principal parts have been lost. Those who control the destinies of the tropical nations of the world therefore have a specially important responsibility for maintaining an outstandingly rich portion of the world's heritage. But, at the same time, it is irresponsible for the international community to expect a nation with few resources to act as unaided custodian for natural treasures that mankind as a whole does not yet fully appreciate.

It is still more irresponsible for agencies charged with providing assistance to developing countries to promote processes that will destroy an irreplaceable resource. Rather, it should be their duty to assist in providing the means for encouraging an appreciation of the resource among the people who own it, for safeguarding the resource and for managing it properly. Such a policy is the only one that can ensure that it will be available for the enrichment of mankind now and in the future, as well as constituting the basis of economic development of greatest and most lasting value to the country concerned.

Essentially, in the humid tropics more than any other region, the ecologist must request a reversal of customary development priorities.

Through appropriate surveys, areas of greatest scientific, educational, aesthetic and potential recreational value should first be identified and protected from any form of use that would lead to their deterioration. However, merely to set them aside or proclaim them as parks or reserves does not even mean that a fair beginning has been made with the task. Such steps have been taken repeatedly throughout the world with little or no effect, except sometimes for briefly postponing the processes of destruction.

What is essential is for a major economic development programme to be launched and directed towards training the scientists, technicians, wardens and other personnel needed for the study, management, protection and development of the areas identified, so that they can fulfil their proper role of serving the interests of the nation and mankind. Support for educational programmes that will build public appreciation and understanding of conserved areas and their resources is a vital part of the process. Provisions for ensuring that the local inhabitants gain materially from the establishment of the parks or reserves are equally important. They must be enlisted among the supporters, custodians and protectors of each area if it is to be successfully maintained.

These objectives require far greater expenditures of development money than have thus far been contemplated in most of the countries concerned. But some indication of the scale of economic benefits alone can be obtained by considering the role of national parks in Kenya and Tanzania as providers of foreign exchange, even though the parks and reserves of these two countries are still only partly developed and their potential for science, education and tourism far from fully realized. Moreover, in some cases, the people in the surrounding areas are not yet supporting or taking a share of responsibility for the development of the reserves, simply because they are too little aware of the benefits, direct and indirect, which could accrue to them.

Once the primary objective as described above is ensured, the second development priority in the humid tropics must clearly be the identification of areas with the highest sustainable potential for agriculture, pastoralism or forest production. Concentration of funds upon intensive development of the best land, much if not most of which is likely to be already in use, should always precede any extensive opening-up of other areas. Improved seeds and agricultural techniques have been successful in recent years in increasing by several-fold the yields from already well-established and productive farm land. One would destroy a great amount of forest and ruin much land seeking to make equivalent gains from the poor soils that support most tropical rainforest—and the gains would be only temporary.

(6) REFERENCES

Aubert, G. (1963). Soil with ferruginous or ferrallitic crusts of tropical regions. *Soil Science,* **95(4)**, 235–42.

Budowski, G. (1965). Distribution of tropical rainforest species in the light of successional process. *Turrialba*, **15(1)**, 40–2.

Dawkins, H. C. (1964). Productivity of tropical forests and their ultimate value to man. In *Ecology of Man in the Tropical Environment*, IUCN, Morges: pp. 178–82.

Dudal, R. (1968). Definitions of soil units for the soil map of the world. *World Soil Resources Reports*, **33**, FAO, Rome. 72 pp.

Edelman, C. H., and Van der Voorde, P. K. J. (1963). Important characteristics of alluvial soils in the tropics. *Soil Science*, **95(4)**, 258–63.

Holdridge, L. R. (1967). Determination of world plant formations from simple climatic data. *Science*, **105**, 367–8.

Holdridge, L. R. (1967). *Life Zone Ecology*. Reviewed edition. Tropical Science Centre, San Jose, Costa Rica.

IUCN (1964). *The Ecology of Man in the Tropical Environment*. IUCN Publications, New Series No. 4, Morges, 355 pp.

Kellog, C. E. (1949). *An Exploratory Study of Soil Groups in the Belgian Congo*. Nat. Inst. of Agronomy for the Belgian Congo, Scientific Series No. 46: pp. 61–71.

Logie, J. P. W., and Jones, G. A. (1968). Land use planning for forestry in Kenya. *East African Agricultural and Forestry Journ.* **33**, Special Issue: 59–62.

McNeil, Mary (1972). Lateritic soils in distinct tropical environments: Southern Sudan and Brazil. In *The Careless Technology*, eds. M. T. Farvar and John P. Milton. Natural History Press, New York.

Mukerjee, H. N. (1963). Determination of nutrient needs of tropical soils. *Soil Science*, **95(4)**, 276–80.

Odum, H. T. ed. (1970). *A Tropical Rain Forest*. Division of Technical Information, US Atomic Energy Commission, Washington DC.

Phillips, J. (1972). Problems in the use of chemical fertilizers. In *The Careless Technology*, Farvar and Milton (eds.), Natural History Press, New York.

Pimlott, Douglas H. (1969). *The Value of Diversity*. Trans. 34th N. American Wildlife and Natural Resources Conference: pp. 265–80. Wildlife Management Institute, Washington.

Procter, J. (1968). Forestry and wildlife land use planning in Tanganyika. *East African Agricultural and Forestry Journ.*, **33**, Special Issue: 63–8.

Richards, P. W. (1952). *The Tropical Rain Forest*. Cambridge University Press, New York, 450 pp.

Richards, P. W. (1961). The types of vegetation of the humid tropics in relation to the soil. *Tropical Soils and Vegetation*. Humid Tropics Research: pp. 15–24. UNESCO, Paris.

Sternberg, H. O'R. (1964). Land and man in the tropics. *Proc. Amer. Acad. Pol. Sci.* **27(4)**, 319–29.

Strahler, Arthur N. (1970). *Introduction to Physical Geography*. 2nd ed., John Wiley & Sons, New York. 457 pp.

Tosi, Joseph A. Jr. (1960). *Zonas de Vida Natural en el Peru*. Memoria Explicativa sobre el Mapa Ecologico de Peru. Bol. Tec. No. 5, Zona Andina, Proy. 39 Program. de Coop. Tecn. Inst. Interamericano de Ciencias Agricolas, Lima.

UNESCO (1959). *Tropical Soils and Vegetation*. Proceedings of the Abidjan Symposium. Humid Tropics Research. UNESCO, Paris. 115 pp.

UNESCO (1963). *A Review of the Natural Resources of the African Continent*. Natural Resources Research, 1. UNESCO, Paris. 437 pp.

UNESCO/FAO (1968). *Conservation and Rational Use of the Environment*. UN ECOSOC, 44th Session, Agenda Item 5(d), 131 pp.

Webb, L. J. (1968). Environmental relationships; the structural types of Australian rain forest vegetation. *Ecology*, **49(2)**, 296–311.

Wright, A. C. S., and Bennema, J. (1965). The soil resources of Latin America. *World Soil Resources Reports*, **18**, FAO, Rome. 115 pp.

CHAPTER 4

Development of Pastoral Lands in Semi-arid and Sub-humid Regions

'In many areas of the world, including both developed and developing nations, semi-arid grazing lands are being effectively destroyed at a rate that appears to be accelerating. The process of conversion of productive land to desert-like wasteland may be observed on most continents. . . . Over wide areas mismanagement of either domestic or wild animals results in overstocking, overgrazing, destruction of the vegetation, loss of soil, and interference with the hydrologic cycle. The damage done may be virtually permanent in its effects. A few owners of domestic animals may receive high short-term profits by the mining of a plant-and-soil resource that could otherwise yield long-term benefits to the many. Knowledge of how to manage such lands is often available to government agencies in the countries involved. Effective application of this knowledge and effective control over land use are almost always lacking.'

From *Conservation and Rational Use of the Environment.* Report submitted by UNESCO and FAO to the Economic and Social Council of the United Nations. Ecosoc 44: E/4458: 12 March 1968.

(1) INTRODUCTION

(a) Importance of Rangelands

In 1967, a panel of experts on the World Food Supply emphasized the importance of effective development of the world's grazing lands for the intensification of animal production. Recognizing the contribution to food supplies derived from the more than three billion domestic livestock that now exist in the world, and from the equal number of domestic fowl, the report put emphasis on increased productivity of domestic and wild animals as a means of improving the nutritional quality of human diets and eliminating protein deficiencies.

Since more than 60 per cent of the agricultural land of the world is considered to be non-arable and suited primarily for grazing, and since domestic animals make effective use of a wide variety of forages, wastes and by-products, the panel believed that animal production could be greatly increased without competing for land useful in the production of plant food. Many countries are largely dependent upon the productivity of their rangelands and animal resources for economic advancement. In most other countries development of pastoral lands and animal resources is a means for improving economic well-being (PSAC, 1967).

(b) Development Problems of Rangelands

There are a few sectors of the environment that have been more badly damaged by man's activities than the grazing lands of the world. There are few fields of economic development characterized by greater or more destructive blunders than those involving attempts to improve the productivity of grazing lands and the domestic animals depending on them. The failures of these development activities have resulted most commonly from attempts to impose new patterns of behaviour upon tradition-bound people, without an adequate programme for education and without sufficient effort to obtain their understanding and wholehearted agreement, and from the consequential inability of governments to control the movements of people and their land use practices. An analysis of these factors lies in the field of anthropology, sociology and politics and is beyond the scope of this volume. Underlying them, however, has been a lack of understanding of the principles of rangeland ecology and range management on the part of the peoples, the development experts and the governments concerned.

The haphazard conversion of productive steppe or savanna land into barren wasteland or desert has been taking place since men first domesticated hooved mammals and began to direct and control the movements of their herds. Many areas have been lost to production entirely, since their potential rate of recovery without major engineering intervention can only be measured on a geological time scale. Much larger areas have had their productivity seriously impaired and continue on a downward trend. Most areas that have been used for the grazing of domestic animals have at one time or another suffered damage from the process. Although the literature is replete with examples, there are fortunately a growing number of instances of the reverse process—the restoration and enhancement of the productivity of previously depleted areas.

Large amounts of development money have been expended by international or bilateral agencies in efforts to increase the economic productivity of rangelands. It would be easy to fill a volume with case histories based entirely on efforts which have failed. In many of these cases a simplistic approach to development has been followed, which failed to take into account the human or ecological factors involved.

(c) Scope of the Present Review

The ecological principles and concepts related to the development of pastoral lands will be presented in the following pages. No attempt is made to be comprehensive since this would require a textbook of range ecology and management. Rather, the ideas considered to be of greatest importance to the development planner have been emphasized. The area concerned lies in mediterranean, subtropical and tropical climatic regions and chiefly consists of savanna, grassland and steppe. It is this type of country, intermediate between desert and the closed forest, that has been of primary value to man in supporting his pastoral industries, although many other uses of such land occur or are contemplated. The term 'rangeland' will be used throughout for these semi-arid and sub-humid pastoral lands.

(2) SOME ECOLOGICAL CHARACTERISTICS OF RANGELAND ECOSYSTEMS

(a) Climatic Factors

The climate of the area under consideration is characterized by extremes. Rainfall is seasonal or sporadic from convectional storms. A long dry season is characteristic. Precipitation normally fluctuates greatly around an annual mean with wet and dry cycles alternating. Several to many wet years will often be followed by a varying number of drought years. As one approaches the desert extreme of rangeland in the dry steppe, rainfall becomes more erratic and less predictable, with a greater tendency to be concentrated in heavy downpours. Dry cycles are more severe. Evaporation and transpiration from vegetation quickly deplete soil moisture, and are accentuated by dry winds.

Rangeland management must be based on the expectation of drought. Some of the most serious land use failures have resulted from the expectation that the above average precipitation of a wet cycle would continue indefinitely. This has occasionally led to ploughing and cultivation of lands that could not permanently support farming and, more generally, to serious overstocking in the course of attempts by livestock owners to ride out what they hope will be a brief drought period. The end product has been the 'dust bowl', depleted lands and spreading deserts.

(b) Soils

There is considerable variation between the soils of semi-arid and sub-humid rangelands. They all tend to be little leached and potentially high in nutrients, but in semi-arid regions many nutrients are locked up in chemical combinations that make them unavailable to plants and mineral imbalances are common, resulting in either excesses or deficiencies. In such conditions the strongly evaporative regime frequently results in high concentrations of salts at the soil surface, creating conditions favourable only to plants with a high degree of tolerance for salinity or alkalinity. In some places trace element deficiencies occur, notably in the wide areas of Australia where it was found that a top-dressing of cobalt salts greatly enhanced range productivity. The influence of the soil's parent material is most marked at the arid extreme and becomes less so where sub-humid conditions prevail. Nevertheless, in view of the great differences in soil components and fertility, any development which involves alterations to the vegetation in the interests of productivity, should always be preceded by a soil survey.

(c) Vegetation

Natural vegetation varies from dominance of woody plants in some savanna and steppe regions to the equally complete dominance of grasses and other herbs. Drought-tolerant plants dominate the arid extreme of the steppe region and extend their range during dry cycles. At the humid extreme, in damper savannas and tall grasslands, plants less tolerant of drought will be found and will likewise extend their area of occupancy during wet cycles but lose ground during dry ones. The vegetation and animal life depending on it are therefore in a constantly shifting dynamic relation to the climate.

Most rangeland areas have vegetation which has been conditioned over thousands of years by the presence of large numbers of wild grazing animals. In Asia, Africa and Europe this has been further modified by many centuries or millenia of use by man and his grazing animals. In the Americas and Australia such use by man is of more recent origin but has commonly brought about marked changes. It may therefore be difficult to judge rangeland potential by examination of existing conditions, which may only reflect the shortcomings of recent management practices. Sites which have been isolated or protected from use by domestic animals need to be identified, so that a better idea can be obtained of the vegetation that could be expected elsewhere under proper management.

(i) Role of Fire

Vegetation reflects a balance between climate, soil, water, animal life, and pressures exerted by man over time. Among the important factors influencing the vegetation of grassland and savanna is fire. In most grassland and savanna fires caused by lightning, volcanism or human action have occurred at one time or another, whenever there has been a sufficiency of dry vegetation and litter. Fires are relatively infrequent in more arid areas, because the accumulation of dry plant material on the soil tends to be less. In the event of a fire, however, the interactions of climate and soil also tend to make recovery very much slower and, with some kinds of vegetation, non-existent. In more humid regions, on the other hand, vegetation is more dense, so that plant litter accumulates more rapidly and fires can occur more frequently. Recovery is also more rapid.

It is dangerous to generalize about the effects of fire, since these vary with timing, in relation to the season of the year, frequency of burning, etc., but it is usually considered that fires favour grass at the expense of woody vegetation. Because of this, fires are often set in attempts to open up rangeland, remove shrubs and make new grass growth more readily accessible to grazing animals. Burning may also have an immediate fertilizing effect upon the surface soil so that the vegetation which springs up after a burn is more palatable, more nutritious and, consequently, more attractive to grazing animals.

It must be pointed out, however, that fires frequently have a reverse effect to that anticipated and can destroy grassland and encourage the invasion of shrubs and trees, if set at the wrong time of year. Because of the adaptation of steppe and savanna vegetation to seasonal drought, there is a period of dormancy during the dry season. Following this period, new growth is put out at the start of the rainy season. To grow vigorously at this time, however, plants must draw upon nutrient reserves stored (in the case of grasses) in the roots and other underground parts. This storage takes place after plant growth declines but before the plant becomes dormant—in other words at the start of the dry season. Fires at this time, which remove the still living leaves of the plant, are damaging to the plant's reserves and when repeated too frequently can kill off grasses. By contrast, burning late in the dry season will usually have less effect on grass. At this time, however, trees often break from dormancy and the new leaves and shoots are particularly vulnerable to fire. Another consideration is that fires may be disruptive to the animal life of the grassland or savanna and, under some circumstances, lead to the disappearance of species.

The value of fire in the management of grassland and savanna has yet to be tested in many parts of the world and requires considerably more research. Nevertheless, certain cautionary principles, which can and should be observed, are worth emphasizing.

Fire should be used with great care, if at all, in semi-arid steppe. In sub-humid savanna, fire is an important tool for management, but the frequency, seasonality and intensity of burning must be carefully studied and the results applied accordingly. In general, as grazing pressure increases, the usefulness of fire as a management tool declines. Burning cannot repair damage done by overgrazing.

(ii) Role of Drought

The vegetation of savanna and grassland is generally adapted to drought. Moderate grazing may favour drought hardiness. Heavy grazing destroys the ability of plants to resist drought, which depends on the storage of nutrients in stems or roots, and crowns. This in turn depends upon the plant having green leafage available during the season of the year when food production takes place. Resistance to drought also depends upon the water storage capacity of the soil, and the barriers to evaporation from the soil interposed by living plant cover and by plant litter. Compaction of the soil due to heavy trampling by grazing domestic animals reduces its capacity to hold water. Destruction of plant cover and litter increases the rate of water loss from the soil and decreases infiltration.

The implications of these principles are, first, that grazing pressure needs to be decreased proportionately more in areas more subject to drought. Secondly, for the reason that the ability of plants and soil to withstand drought depends in part on their condition at the end of the wet season, one way to improve this condition, both in respect of plant stamina and soil structure, is always to leave a reserve of standing forage and litter at the end of a growing season. In the western United States, for example, it has been estimated that from one third to two thirds of above-ground plant material should be left ungrazed. A rough rule of thumb, which must be modified according to local conditions, is to take half and leave half.

It may be important to graze less desirable or relatively unpalatable vegetation heavily early in the growing season to prevent it from spreading. Grass that becomes tall, coarse and unpalatable late in the season must be managed differently from grass that retains its palatability through the year. Local rules must be based on local conditions, and on studies by competent experts. Once established, however, the rules must be followed or rangeland will deteriorate.

(d) Water Relationships of Soil and Vegetation

Although the vegetation of semi-arid rangelands is adapted to survive drought, this does not mean that it will not be wasteful of water when water is available. Indeed, many desert and steppe shrubs are notoriously wasteful of soil water. Although the following generalizations do not always apply to local situations, they have been found to be widely applicable.

An increase in woody vegetation at the expense of grasses and broad-leaved herbs (forbs) will cause a much greater drain upon soil water resources. Following overgrazing, which weakens grasses and permits invasion by less palatable shrubs, springs often cease to flow and streams become intermittent and undependable. Removal of the woody vegetation and restoration of the grassland cover often results in an increase in the output of springs and restoration of normal levels of stream flow.

As previously noted, heavy trampling by livestock can cause soil compaction, and consequently decreased water-storage capacity and increased run-off. This eventually leads to sudden 'flash floods' after heavy rain, followed by periods when neither springs nor streams flow. Reduction of grazing pressure and restoration of soil permeability will usually bring about an improvement in the amount and reliability of flow.

Heavy grazing pressure, through removing vegetation and exposing soil to the impact of rainfall, also increases erosion, among the consequences of which will be silting up of channels, stock-watering tanks and reservoirs. A far better water supply is therefore likely to be ensured by proper management of livestock, vegetation and soil than through expensive construction of elaborate water storage structures. On poorly managed lands the latter are likely to have a short life.

(e) Productivity of Grassland and Savanna

Looked at from the viewpoint of production of green plants, grasslands and savannas are intermediate between highly productive forest areas and the low productivity of deserts. However, in the production of protein of value to man, grasslands and savannas are superior to all other natural vegetation types. In part, this is a reflection of the nature of the soil: soils of less humid regions are not leached to the extent that is characteristic of most forest soils; they retain a high amount of calcium, a good ratio between calcium and phosphorus and in general a proper balance of all of those elements necessary for abundant protein production. The better forage species of grasslands will commonly have a protein content of 13 per cent dry weight or more, equivalent to that found in well-cured alfalfa (lucerne) hay. In part, however, the high production of animal protein is related to the fact that the principal plants are within reach of grazing and browsing animals. When grasslands are converted to cereal grain production, they remain high as protein producers, and the crops will average a higher percentage of protein than those produced on forest soils (discounting the effects of fertilizers).

The former abundance of wild grazing and browsing animals in grasslands and savannas is almost legendary. The great herds of American bison, of Eurasian saiga and gazelles, of Indian antelope, deer, buffalo and elephant, or of the array of large African mammals, reflect the primeval productivity of the world's grasslands.

Productivity varies within the grassland and savanna region. Vegetative production is higher near the grassland–forest boundary and falls off towards the desert edge. Animal protein production by grazing mammals probably reaches a maximum in an intermediate zone between sub-humid and semi-arid grazing lands. The better watered grasslands, where soils are suitable, may be readily converted to permanent pastures or cultivated cropland. Towards the arid extreme, except where irrigation is possible, conversion from natural vegetation is not economically practical; attempts to grow crops are accompanied by frequent failures and accompanying erosion and soil damage.

In many of these more difficult areas, there is evidence that production from native wild animals, adjusted to the climate and able to make use of the wide range of natural vegetation, will exceed production from domestic livestock. Intensive management could change the balance in favour of domestic livestock—for which the management rules are better known. However, equal care and attention given to wild animal populations might also lead to far higher yields. In either case, intensive management is rarely economically feasible, and the question is which group of animals can make the best use of natural vegetation under existing conditions. In answering it, the multiple values of wild animal populations need be balanced against the narrower range of values for domestic livestock.

Some of the greatest differences in rangeland productivity will be found not between the sub-humid and semi-arid ranges but between well managed and poorly managed ranges in the same climatic and soil region. Poor management can reduce the yield of a fertile and well watered range to a lower level than that found in a well managed natural desert.

(f) Stability of Grassland and Savanna

Grasslands and savannas occupy a tension zone between forests and woodlands on the one side and desert on the other. On the arid side, disturbance of grassland by overgrazing or prolonged periods of drought leads to replacement of grassland by desert vegetation. On the humid side, invasion of grasslands by woody vegetation follows on overgrazing or the misuse of fire. Despite this, it is not unrealistic to consider grassland vegetation as a relatively stable, if not climax, plant growth in many parts of the world and to manage it for continued production of forage and of the animal life dependent upon it.

Most grasslands have evolved along with the complex of hooved grazing and browsing animals that depend upon them. Grassland plants in most parts of the world are adapted to grazing and browsing and are able to continue growth and reproduction in its presence. In some areas, however, notably Australia and New Zealand, the native grassland developed in the absence of hooved herbivores and in consequence was fragile and susceptible to change when domestic and wild ungulates were introduced. However, a new and relatively stable balance has eventually been achieved and it is now feasible in both countries to manage grasslands and savannas for sustained production of domestic livestock.

(i) Ecological Succession

Stable or climax grasslands are dominated by perennial grasses. These are of the tussock or bunch-grass form, growing in compact clumps, or are tall grasses or shorter species of the sod-forming type which spread by rhizomes or stolons, forming a continuous cover over the ground. Under the pressure of grazing animals there will be a shift in the composition of the grassland. Those species preferred by animals, usually of high palatability, will undergo maximum grazing pressure and those least resistant to grazing may disappear. Species best able to stand grazing or those which are unpalatable to grazing animals will increase. Under excessive grazing pressure, virtually all palatable species will disappear. The ground will then be occupied by the least palatable species. These may be tall, coarse grasses, such as *Cymbopogon afronardus* in East Africa, which is exceedingly difficult to eradicate once it takes over. It does have an advantage, however, in that it protects the soil from erosion and gradually builds up a layer of litter and humus. Less desirable is the situation in which the ground is occupied by weedy annuals which spring up from seed each year, but give little soil cover. Under very heavy pressure of grazing and trampling all soil cover may be destroyed and the land exposed to baking and erosion.

Grasses are frequently classified, for range management purposes, as increasers, decreasers and invaders (*Dyksterhuis, 1949*). *Decreasers are the climax perennials which cannot tolerate grazing pressure or which are most palatable to livestock and will diminish and eventually disappear under heavy grazing use. Increasers are part of the perennial climax but are better able to hold up under grazing or less favoured by grazing animals. These gain ground under moderate grazing pressure. Under heavy grazing, however, even some or a majority of the increasers are grazed out and the ground is then occupied by invaders and by the least palatable and most grazing-resistant increasers. Invaders are weedy species that are normally restricted to small, heavily disturbed areas when the grassland is in good condition, but which are capable of spreading everywhere when grazing is excessive. A range in good condition will contain a mixture of increasers and decreasers.*

Invaders, both in tropical savanna and in semi-arid steppe, usually include a variety of woody plants. Although some shrubs and trees are highly palatable to grazing animals, the forms that tend to occupy the ground under heavy grazing use are those which are least palatable and commonly are thorny forms avoided by browsing species.

Deterioration of grasslands under heavy grazing is not solely caused by the actual grazing but also by trampling and the soil compaction which occurs where large numbers of domestic livestock are concentrated. This decreases soil aeration and permeability to water, so that much of the rain water runs off. The soil then becomes dry and less capable of supporting plant growth, even though the total amount of rainfall is adequate. With exposure of soil to wind and water erosion is accelerated and, in the end, only a rather sterile substrate or even bare rock may be left.

(ii) Recovery Rates and Thresholds

In general, the rate of recovery of a grassland from grazing varies directly with grazing pressure and the degree of subsequent disturbance. Grasslands that are moderately grazed recover quickly. Under heavier grazing pressure or more severe disturbance, recovery following the removal of grazing is slowed down. If the more palatable species are completely killed out it may take many years before they can recolonize the area. If they are destroyed extensively over a wide area, they may be unable to do so at all, for a new balance of species may have become established. This happens particularly when invading weedy species or brush have occupied the ground completely. The only management solution then possible is to destroy the weedy plants and reseed to the better forage species.

There are obvious thresholds beyond which further grazing or other disturbance does not allow quick recovery. One is the threshold that permits invasion by vigorous and highly competitive weeds or shrubs. A far more dangerous one, however, is that at which soil damage begins to take place at an accelerating rate. On many rangelands a point will be reached where sufficient bare ground has been created to allow rapid erosion. If this occurs, deterioration of the site may continue even with complete protection, since the rate of soil loss will exceed the rate at which grasses or other rangeland plants can invade and hold the ground. Such deterioration can be checked by immediate attention to erosion control and reseeding with suitable plants. In the absence of such management, however, the site will deteriorate to a point where eroding and stabilizing forces can reach some new balance. Commonly, the usefulness of the area will be largely destroyed before this point is reached and recovery will be extremely slow.

(iii) Desertification

Most of the deserts which are spreading at the present time are doing so because of abuse of the land around their borders rather than through any changes in climate (Figure 9). Destruction of native plant cover by excessive grazing, by ill-considered attempts at cultivation and farming or by other disturbance, will usually begin the process of desertification. As noted earlier, compaction and damage to the soil can create desertlike

Figure 9. More deserts spread through abuse of the land around their borders rather than through any changes in climate. Desertification in Pakistan. (Photograph by C. D. W. Savage: courtesy World Wildlife Fund).

conditions even where rainfall is adequate for plant growth. Erosion can carry the process still further, removing the soil and leaving a substrate unable to support much plant growth. Drought-resistant, hardy, desert plants will then move into areas that previously supported steppe grasslands or savannas. The aspect of the country will change to resemble a climatic desert. Drifting sand, blowing dust, dry watercourses, barren gullies and other desertlike conditions will prevail. Despite this, the regional climate will retain the capability for supporting productive vegetation. Management must involve, initially, destocking, or other removal of the causes contributing to desertification and, secondly, control of erosion by appropriate biological or engineering techniques. Restoration of vegetation by reseeding and replanting on the better sites with species best adapted to the existing conditions will be the next need. They will commonly be species that are drought hardy or otherwise able to tolerate the more extreme conditions of the area. As these build up soil and modify the local environment it will be possible to replace them with more desirable plants.

The entire process of reclaiming lands which are undergoing desertification is dependent, however, on the ability to control use of the land. If the grazing or other disturbing processes that originally contributed to desertification cannot be excluded, there is little point in attempting reclamation.

(g) Influence of Grazing Animals

As noted earlier, most grasslands have evolved under the impact of grazing animals and the vegetation is as much in balance with their presence as it is with the climate, soils and other factors of the environment. In the absence of grazing a different vegetation would prevail, in the same way that, in the presence of excessive grazing, the vegetation alters its general state and composition in a less desirable direction. Most if not all plants can tolerate some degree of animal use. Their response varies with the intensity and the timing of that use.

Range grasses usually have three periods in each year when they are most vulnerable to grazing pressure. The first is at the start of the growing season when the plant is dependent upon reserves of nutrients stored in the roots or root crown. Grazing of the newly sprouting vegetation can prevent the plant from establishing enough photosynthetic surface to manufacture the food materials it requires. When reserves are exhausted the plant will die. A second period is after the main growth of the year is completed, when the plant is developing and maturing a seed crop. Use at this time can prevent seed from being set or cast and thus endanger reproduction. A final period is towards the end of the growing season when the plant is storing reserves for next year's growth. Heavy use of the grasses at any of these three stages is more likely to be injurious than at other times of the year, and various systems of rotational and deferred grazing have been devised to provide protection accordingly.

The growing buds of grasses are located near the base of the plant. Grazing of the top leafage is therefore seldom injurious, except at the times noted above. Close grazing, however, which removes or exposes the growing tissues, is generally injurious. Nevertheless there are great differences in the ability of various species to withstand grazing. Grasses with an open, loose growth habit are more likely to be injured by grazing than a compact bunch or tussock type. Grasses which form a sod, reproducing by rhizomes or stolons, can usually better withstand grazing than the taller grasses. Certain types of grassland, dominated for example by sod-formers such as *Cynodon* or *Bouteloua*, which are highly desirable for forage, thrive in the presence of moderately heavy grazing and disappear through replacement by other species in the absence of grazing. Annual grasses, after their seeds are mature, are not injured by any amount of grazing since they die back completely in any event. But conditions for the germination of seeds and growth of seedlings are improved if grazing is not too close and a cover of dry stems and litter is left on the ground.

If animals move about and are not allowed to concentrate in any one area for long, they tend to graze selectively, picking out those species or portions of plants that they prefer but doing little overall damage to vegeta-

tion. Animals that are held for prolonged periods in limited areas can have an adverse effect, through overgrazing or trampling, on even the most resistant species.

Wild grazing animals, when unconfined, are inclined to graze selectively and keep on the move, thus not exerting continuous pressure in any one area. A variety of different species such as are found in the savannas of East Africa will make use of a wide range of different plants, each species tending to have a different preference. The presence of a great number of wild species, each adapted to its own place in the savanna, results in a relatively uniform use on the entire range of plants in the vegetation. Selective pressure thus does not favour the less desirable rangeland plants and range vegetation holds up well under what amounts to very heavy total grazing pressure. An equal number of any one species of grazing animal would do far more damage than the combination of many species.

Many wild species are adapted to go for long periods without drinking water and, in consequence, do not confine their use to areas within easy reach of watering points (Figure 10). Eland, oryx and addax are three African species which can occupy grazing lands far removed from permanent water sources. By contrast, most domestic species need water much more regularly and their grazing is confined to areas that are not too far from watering points.

Figure 10. Many wild species can go for long periods without drinking. A potential candidate for domestication is the addax which inhabits sandy dry areas of the Sahara. (Photograph by F. Edmond-Blanc).

Traditional nomadic grazing of steppe and savanna lands was based on a grazing system that most closely resembled that of wild grazing animals. Nomads kept their animals on the move permitting selective grazing, but usually not allowing a heavy concentration of animals to remain in any one place. Although routes of movement were determined by the location of suitable watering points, livestock were not kept for prolonged periods in the vicinity of any one water hole. Local concentrations did take place during the winter season of mediterranean or subtropical lands, and in the dry season in the tropics, but these were periods when the vegetation was least vulnerable to the effects of heavy grazing. In general, therefore, nomadism was well adapted ecologically to the conditions prevailing in areas occupied by nomadic peoples.

The impact of development on such areas has tended to be unfavourable. Initially, the effects of introducing better provision for maintaining the health of people and animals removed constraints that had previously prevented great increases in numbers. The elimination of intertribal warfare removed another limitation to population growth. The desire on the part of governments to confine and sedentarize nomads discouraged free movements of peoples and their livestock and contributed to the concentrations of grazing animals where they would be most likely to damage the vegetation on which they depended. The education and training needed to permit the nomads to adapt to a different way of living has not accompanied the constraints on their former way of life. The result has been the semi-nomadic and semi-sedentarized patterns of land use which contribute most strongly to the destruction of rangelands and the spread of deserts. Some of the considerations affecting nomadism and pastoral peoples in the modern world have been well expressed by Derek Tribe *et al.* (1970) in a paper presented to the UNESCO Conference on Man and the Biosphere held in Paris in 1968:

'In areas of arid or semi-arid land, covered with dry bush or steppe, people are often pastoralists from necessity and it may be undesirable to alter substantially indigenous systems of land use based on nomadism. In conditions of erratic and scanty rainfall, a nomadic habit can make the most efficient use of water, and the ephemeral green vegetation where and when it is available. Conversely, to turn the nomad into a settled rancher (particularly on a small scale) by providing additional water supplies, so that instead of ranging over a wide area he is confined with his herd to a limited locality, is likely to lead to widespread land deterioration. It is often better to seek the solution by combining nomadism with further development in irrigated or high rainfall areas.

'Efforts to raise the imaginative horizons of pastoral communities must concentrate upon loosening their traditional regard for livestock.... When he begins to regard livestock in terms of their monetary equivalent, the pastoralist must at once begin to appreciate quality in 'differential price values'. Only when this change in attitude has been achieved is he likely to cooperate in improvement schemes involving destocking or a rational use of water and grazing resources.

'There is no point in introducing a controlled grazing scheme in an area in which livestock owners are reluctant to sell their cattle, and insist on grazing their land according to tradition.'

(3) ASSESSING STATUS AND CHANGE IN RANGELAND ECOSYSTEMS

(a) Introduction

Existing patterns of grazing use of most of the world's rangelands are unsatisfactory and contribute both to growing human suffering and dissatisfaction as well as destruction of natural resources. However, there are few areas where things have gone beyond the point of repair. Rather, in most rangelands of the world, a relatively small expenditure of development money, wisely applied, can go far towards restoring healthy and productive vegetation for the support of an abundant animal life and for the enrichment of man. Essential to this process, however, is the ability to recognize the existing status and trend of rangelands in relation to the potential for the site, identify the factors causing damage, and institute a total management programme, with the support of the local people, to eliminate those factors and develop the highest potential of the area.

Development planning must be based upon a knowledge of existing conditions in the area to be developed. For rangelands this implies, among other things, a knowledge of the actual as compared with the potential condition of the vegetation of the particular site, a knowledge of the factors that have resulted in the vegetation reaching its present condition, and a knowledge of the direction and rate of change under existing conditions of management. Changes brought about through development must further be evaluated, if one is to gauge progress and the gains or losses in animal and human welfare resulting from the development efforts.

Insufficient attention to these requirements has been an underlying cause of failure in many development projects. It is fortunate, therefore, that techniques for the rapid assessment of the condition and trend of rangelands have been developed. While requiring continual refinement and improvement to adapt them to local conditions, they are nevertheless widely useful. In the hands of a trained range ecologist or manager, they can be developed into working techniques which may then be applied extensively by less highly trained individuals. Their use moreover permits people affected by development proposals to judge for themselves the statements and predictions of experts.

In subsection (b) which follows, the factors involved in judging the condition and trend of rangeland are discussed. It must be stressed that condition and trend reflect rangeland management, or its absence. The greatest single factor contributing to the decline in the productive potential of rangeland is mismanagement of livestock. Overgrazing or improper grazing will cause any area of rangeland to deteriorate. Absence of grazing, also, may cause rangeland to change in directions less desirable to those who intend to use the land for forage production. The number, spatial distribution and kinds of animals, and the seasonal pattern of the grazing must all be controlled if rangelands are to be maintained in productive condition for higher yields and profits.

(b) Techniques for Assessing Condition and Trend

(i) Range Condition

This term covers the assessment of the status of a particular area of range or pastureland in relation to its potential. *An area rated in excellent condition will be one in which forage production, both in quantity and quality, approaches an optimum for the site in that soils and vegetation are stable and conditions are favourable for continued high productivity. In some areas in which grassland is considered climax, range condition is measured as the degree of departure from the stable, climax mixture of grasses and broad-leaved plants (forbs). However, on many areas of range the most highly productive condition, measured as quality of forage for livestock and quantity of high quality forage produced, will not be climax but some successional state, which may be maintained by grazing, fire or other factors.*

Condition is not to be confused with *site quality*. A range in excellent condition in an arid area with shallow soils will produce far less forage than an area in equally good condition in a more humid climate with deep soils.

Condition classes are commonly classified as 'Excellent', 'Good', 'Fair', 'Poor', with sometimes a 'Very Poor' condition also recognized. Assessment of range condition depends upon ability to recognize and classify the various species of plants, and to recognize their successional position in the vegetation and their value to the class of animals being considered. Perhaps even more, it depends on ability to recognize the symptoms of site deterioration as revealed by the amount of ground covered by various types of vegetation (e.g. trees, shrubs, perennial grasses, annual grasses, forbs, etc.), the amount of bare ground and of dead plant material (litter), and by evidences of erosion.

Techniques for rapidly assessing condition have been described by Riney (1956, 1963), Dasmann (1948, 1951), and others. Many of these techniques have been developed for specific regions; however, if the species listings are ignored, the general principles are applicable to all areas. Assessment of range condition requires training in range ecology. However, the techniques and procedure, once developed by a range ecologist, may readily be used by students, technicians, livestock managers and all other interested individuals. Techniques for the quick classification of large areas of rangeland through the use of aerial photography are now under development.

(ii) Range Trend

A companion measurement that must accompany the assessment of range condition is that concerning range trend. This states, in effect, whether a particular area of range or pastureland is deteriorating, improving or holding its own, under current conditions of use and management. Thus, an area of range classified in 'Excellent' condition but noted to have a *deteriorating trend* is an area for management concern. An area in 'Fair' condition, but noted to have an *improving trend* is evidence of good recent management.

Measurements of range trend are based upon an assessment of evidence of changes in vegetation—whether or not seedlings are becoming established and of which species or class of vegetation, and of changes in soil condition —whether or not erosion is active and accelerating, or whether or not evidence of past erosion is being 'healed over' by plant growth.

For example, an area in which the tussock or bunch form of perennial grass represented the desired type of vegetation, in which old clumps of these grasses were isolated on soil 'pedestals' (indicating the eroding away of soil between clumps), in which there were no seedlings, in which all gullies and rills were sharp edged and showed signs of recent active erosion, in which wind-blown soil was piled on the windward side of grass clumps, rocks or other physical barriers, in which dunes were forming, or in which stones and pebbles normally covered by soil were exposed on the surface (erosion pavement), would be considered as *deteriorating*. In most tropical savanna rangelands the spread of woody vegetation and the spread of bare ground are two of the clearest indicators of such deterioration. The procedure and techniques for assessing trend need to be developed for any region by a competent range ecologist. However, once developed, the guidelines, as in the case of techniques for assessing range condition, can be followed by any conscientious person.

(iii) Key Areas and Key Species

It is seldom necessary to examine range condition and trend over exten-

sive areas of rangeland. Fortunately, in most regions it is possible to identify the so-called key areas on which livestock, wild animals or man himself must depend during the critical season of the year. The condition and trend of these areas determine the long-term survival of the animal populations that depend upon them. By contrast, the condition and trend of other areas may be relatively unimportant, and in some instances deceptive.

Key areas in regions where winter cold and snowfall limit the movement of animals, the accessibility of forage or the condition of that forage, are to be found in the wintering ranges of the animals. For most species these areas are small in extent, relative to the total range available, and easily identified by either the presence of animals or evidence of their past presence. Here the animals must survive during the most limiting period of the year at a time when they are subject to the most severe climatic stress. In tropical climates, key areas will be found most frequently in areas near permanent water. Here those animals that require drinking water must find their food supply during the dry season. The extent of the area involved will vary with the species. Horses and zebras may travel many miles from water, returning to drink at intervals of as long as three days. Cattle will have a shorter radius of movement; giraffes longer.

Regardless of the species of animal involved it is the condition of the range within the area that must support the animal during the critical season of the year that is all important. It is of minor importance that excellent forage exists in inaccessible areas or areas that cannot be reached by the animal without risk of dying from thirst.

In assessing key areas in dry climates, it is the total condition of the area used that must be assessed. Areas in the immediate vicinity of water will normally show very heavy use even though the entire area available to the animals is only moderately used.

Within key areas it is usually possible to identify those key species of plants upon which the animals depend most heavily for food during the critical period of the year. Assessment of the condition and trend of these key species may save the labour of attempting total assessment of the entire complex of vegetation.

Nevertheless, an assessment of the overall status of range condition and trend within the key area remains essential for estimating the likelihood that the key species will continue in a healthy productive condition. This is because some species which are considered key species for the survival of animals at a particular time, may themselves be indicators of a deteriorating condition for the area as a whole (e.g. certain species of *Erodium* are important forage plants for sheep, but when present in abundance can be evidence of a range about to undergo further deterioration to a highly unproductive condition).

Once again, although the identification and initial assessment of key areas and species requires the skills of a competent range ecologist, continued monitoring can be carried out by personnel who can be taught quickly the relatively simple techniques involved.

(iv) Other Biological Indicators: Evidence from Changes in Species Composition of Vegetation

As noted earlier, heavy grazing, associated with a deteriorating trend in rangelands, will cause a decrease in those species most palatable to grazing animals and least resistant to grazing. To determine the direction of range trend it is necessary to observe the various age classes of the species occupying the rangeland. If the preferred species are represented only by older individuals of declining vigour, and younger plants are of less desirable species, a deteriorating trend is apparent. Similarly, a depleted range that shows great numbers of seedlings or young plants of better forage species, is considered to be improving.

Obviously, ability to recognize the characteristic species of the various successional stages of the rangeland under examination is a prerequisite for using these criteria.

(v) Evidence from Changes in the Height, Vigour or Health of Plants

Disappearance of a desirable species of plant from a rangeland as a result of range deterioration will often be preceded by changes in height, vigour or health. These are not always readily detectable unless other areas of well-managed or undisturbed rangeland are available for comparison, since changes in the appearance and productivity of range plants also accompany year to year variations in rainfall or other climatic conditions.

In general, the response of a plant to excessive browsing or the deterioration of its site through soil exposure, compaction, erosion, etc., will include a loss of resistance to various plant diseases, a decrease in height and leafage of individual grass plants or forbs, a dying back of portions of tussocks or bunch grasses, reduced sprout growth in shrubs, increased evidence of insect and disease damage to shrubs, or dying back of twigs and branches, and often a yellowed, unhealthy appearance of the foliage. Since such changes can result from a variety of different causes not always related to the degree of grazing pressure or soil damage, they must be evaluated with considerable care, and sometimes with the assistance of experts in the physiology or pathology of the species involved.

It is common folklore among pastoralists and hunters that if animals appear to be in good condition their habitat is necessarily in good condition. In fact animals can often remain fat and healthy during a period when their habitat is undergoing severe and rapid deterioration. Only when late stages of deterioration are reached will the animals fall off in condition. When this occurs the change may happen both rapidly and drastically and the opportunity to repair the damage to the habitat easily and quickly will have been lost.

Despite the above qualification, it is still possible to use animals as indicators of habitat conditions. Their general state of health is more obvious to the untrained observer than are the evidences of change in vegetation and soil, and may sometimes provide the first indication that all is not well.

(vi) Evidence from the Decline in Condition of Grazing Livestock and Herbivores

Lack of body fat, appearance of bony structures (ribs and pelvic girdle) normally concealed by body tissues, loss of glossiness of coat, patchiness or roughness of coat, are all signs of poor body condition among livestock. Internally, loss of fat from the body cavity and bone marrow will be evidence in carcasses. Even the skeletal remains of animals long dead may be assessed for condition by examining the evidence of fat within the bone marrow cavity. Techniques for such assessment have been developed by Riney (1955), for large wild mammals, as well as by others (Giles, 1969).

In deer, a decline in antler size and weight and an increased frequency of malformed antlers accompanies a decline in level of nutrition. Increase in the number and visibility of external parasites commonly accompanies a loss in condition along with an increased prevalence of internal parasites, and a greater incidence of disease.

Reproductive ability is often changed in response to body condition. A decreased production of young will accompany a fall-off in nutrition in most species—fewer young per litter and fewer litters. Under severe nutritional stress many species will cease breeding until such time as conditions improve.

Young animals are sensitive to nutritional stress since their nutritional demands, relative to their size, are greater because of rapid growth. A poorer survival rate of young often precedes an actual decline in birth rate.

(vii) Evidence from Changes in Species Composition of Wildlife

As indicated in subsection (iv) above, a decline in the condition of a a rangeland is commonly reflected in less diversity in the species of plants present over any extensive area. The change in plant species composition commonly brings a change in animal species composition. Among domestic livestock, for example, horses are relatively exacting in their habitat requirements. Goats are highly adaptable. A range no longer capable of supporting horses may well support, for a time, great numbers of goats which can feed on shrubs, trees, and a wide variety of grasses and forbs. Inevitably, the presence of great numbers of goats will result in a further and serious decline in rangeland condition, although goats are not necessarily undesirable and, particularly on shrub-covered ranges, can make a valuable contribution to the local economy if they are correctly managed.

Similar changes may be expected among wild grazing animals. Certain species are quick to disappear as range condition declines. Thus, Savory (1969) noted the disappearance of bushbuck from heavily grazed and browsed riverine forest in Africa. Sable and roan antelope may show a similar sensitivity to changes in habitat conditions. By contrast, an overabundance of elephant, buffalo, zebra or impala may indicate a serious degree of rangeland depletion (Dasmann and Mossman, 1962; Savory, 1969).

The relationship of rodents and rabbits to deterioration of rangeland has been well studied in the western United States and elsewhere. Although an overabundance of these so-called pests is often considered a cause of range depletion, many ecological studies suggest that it is instead a consequence of rangeland depletion by livestock grazing. Linsdale (1946) noted in California the disappearance of jackrabbits (*Lepus*) and ground squirrels (*Citellus*) from a California oak savanna following protection from livestock grazing. Koford (1958) has observed the inability of the prairie dog (*Cynomys*) in the middle western United States to invade areas which had been protected from livestock grazing or similar disturbance. Reynolds (1958) and others have noted the relationship between the abundance of kangaroo rats (*Dipodomys*) and the condition of the range as influenced by domestic stock. Similarly, the restriction of the European rabbit to areas affected by previous disturbance of the vegetation or other causes has been noted in England and Australia.

Although many species of rodents and rabbits have been considered 'animal weeds' which infest rangelands, they may better be considered as animal indicators of depletion brought about by other causes. It must be recognized, however, that many of these species are fully capable of keeping a rangeland in a depleted condition once they are established. Range improvement may require their removal.

The rapid improvement in some areas of European and Australian grassland following the removal of rabbits by the introduced disease *Myxomatosis* is illustrative of this principle (Ratcliffe, 1959). Although most work on animal indicators of rangeland depletion has been carried out in North America, similar results may be expected from studies on other continents and relationships may quickly be established by rangeland ecologists.

(4) IMPROVING RANGE CONDITION THROUGH DEVELOPMENT

(a) Controlling Livestock Numbers

Unless control can be obtained over the number of animals grazing or browsing in an area very little can be done to improve the condition of the rangeland (Figure 11). If numbers of animals are uncontrolled any improvements that can possibly be brought about will have only the most temporary effects, since grazing animals will concentrate in or around them to create as much or greater damage than existed before.

Control may be brought about by an agreement with the graziers to sell off or otherwise remove excess stock, or it may involve only an agreement to hold livestock off the area to be improved through herding or fencing. Achieving such an agreement, however, often involves programmes of public education, training and all-round salesmanship. For reseeding, removal of undesirable vegetation or many other forms of range improvement to take effect, there must usually be a period when all livestock are removed, and for other types of range improvements there must be an

Figure 11. Overstocking and excessive grazing in India. There is no point in attempting range improvement where livestock numbers cannot be controlled. (Photograph by E. P. Gee: courtesy World Wildlife Fund).

agreement to hold animal numbers below the actual carrying capacity of the area involved.

With wild grazing animals it is sometimes exceedingly difficult to control numbers or reduce overstocking without an intensive hunting effort. Methods of controlling wild animals through shooting, however, involve a great number of risks and are sometimes difficult to execute. Commonly, the more vulnerable animals—such as the herding species that prefer open terrain—can be quickly reduced, whereas the more solitary species that prefer dense vegetative cover may be little affected by most hunting. Unless hunting is carried out under careful supervision of qualified wildlife biologists and managers it may well produce results far different from those intended. Furthermore, preceding any mass reduction in wildlife, provision needs to be made to take full advantage of the scientific information to be obtained from the animals killed, and to avoid wastage of meat or other valuable animal products.

(b) Grazing Management—Controlling Distribution

Range damage often occurs not because an area is grossly overstocked, but because the distribution of animals is uncontrolled. A range that could potentially support 1,000 animals may yet show serious damage with a stock of 100, if they are allowed to concentrate in the wrong place at the wrong time. Frequently, the condition of a range will show marked improvement as one moves out from an area preferred by the animals involved. Much of the area may be little used, although the area around the preferred location will be grazed excessively.

Control of animal distribution can be influenced in a variety of ways: by changing the number of distribution of watering points; by providing sources of salt or other desirable minerals in the appropriate localities; by the location of areas in which supplementary feeding will be conducted; or, where manpower is available, through active herding. The most effective and permanent technique for controlling distribution is through fencing, although in some circumstances this can be excessively expensive, as it usually is when wild grazing animals are involved. Some wild animals can be excluded by normal stock-proof fences but many will break down or jump over any but specially constructed, and usually expensive, barriers.

As noted earlier, many range grasses need special protection at critical times of their life cycles. Grazing can be heavier at other seasons, but must be lessened during these critical periods. To meet this requirement various systems of rotation and deferred grazing have been devised that will concentrate use seasonally to allow certain pastures to recover and improve in condition and to give all pastures some degree of seasonal relief. When properly managed, such systems can produce marked improvement in range condition and total forage yield without a change in the overall level of stocking. The system best suited to a particular rangeland must be worked out by a range manager experienced with the area to be managed.

Controlled use of fire is a means of shifting concentrations of both livestock and wild animals in order to allow certain areas to recover from overuse. Usually grazing animals will be attracted to the newly sprouting forage on a burned area and will concentrate their feeding activity on new burns, thus allowing other areas to recover. As previously emphasized, however, fire is an uncertain management tool unless its use is based on study and experience of the conditions under which it is to be employed. Differences in the intensity, frequency and seasonality of burning can produce results that vary greatly and may be opposite to what was intended.

(c) Control of Woody Vegetation and Undesirable Plants

Fire can be used under some circumstances to remove undesirable woody vegetation and allow for its replacement with more desirable species. Once again, the need for skill and care based on appropriate local studies and experience must be stressed. Furthermore, there are many situations where fire will not be effective. Wooded areas from which ground cover or deep litter is absent may not support a fire and overgrazed grasslands usually cannot be burned.

Mowing, crushing and cutting have been used to remove woody plants, sometimes accompanied by the use of fire to eliminate the plant debris which is thus accumulated, and restore its chemical nutrients to the soil. Many less desirable woody species, however, will sprout vigorously after cutting, crushing or burning. Follow-up management will then be essential or the scrub that sprouts up will be denser than that which preceded the clearing operation.

Certain undesirable range plants can be controlled by concentrating animal use in areas where they occur at the time of year when they are most vulnerable to grazing. Essentially, this is a process for deliberately overgrazing an area in order to destroy a particular type of vegetation. Unless a follow-up treatment of the area is provided, however, usually involving reseeding with desirable plants, the effects of such practices can be adverse. Worse weeds will replace those destroyed as range condition continues to deteriorate.

Some plants are poisonous to domestic livestock and at times create serious problems for the grazier. Wild animals appear not to be affected except under unusual circumstances, and there are considerable differences among domestic breeds, both in tendency to feed upon poisonous plants and in the degree to which they are affected by various plant toxins.

Problems of control vary with the successional stage of the plants concerned and their ability to compete with other species. Those which typically appear at an early stage in the succession may often be eliminated by programmes that will bring overall range improvement. Others, however, hold their own even on well-managed or climax rangelands and require special measures for eradication. Species which are also poisonous to man will require special care. It is not possible to generalize about methods for control and expert advice will usually be required. The use of chemical herbicides is discussed in subsection (e) below.

(d) Improving Water Development

'Experience has repeatedly shown that water development without land reform, grazing control and cooperation from livestock producers leads

Figure 12. Water development without land reform or grazing control has led to wind erosion and desertification, as shown here, as well as most desert edge countries. (Photograph by National Parks Board, South Africa).

rapidly to the destruction of grass cover by serious overgrazing, bush encroachment and soil erosion (Figure 12). The damage that can be done in five or ten years may take many decades to repair' (Tribe *et al.*, 1970).

The location of drinking-water governs the distribution and movements of most species of grazing animals during the dry seasons of the year. If suitable watering points are well distributed over a rangeland, range use will be more even and local concentrations of animals are less likely to develop. Where drinking water is not available, only wild species able to exist without it will be able to utilize the forage.

In recognition of this principle, it has long been apparent that development of new watering-points in dry areas is a means of improving the carrying capacity of the land and achieving a better distribution of animals. It is unfortunate, therefore, that such developments have too often resulted in the destruction of large areas of rangeland and the conversion of previously productive regions into deserts.

The spacing of watering-points must take into account the frequency with which the animals concerned drink and their radius of movement in the course of feeding away from water. To achieve utilization of the whole of a particular area of rangeland, watering-points should not be further apart than twice the average radius of movement and, for more even utilization of the range, they should be closer together than that, in order to eliminate the tendency of animals to over-use areas in the vicinity of any watering-point.

Water development should not be attempted in a region in which overgrazing already exists and in which no control can be exercised over the numbers or distribution of grazing animals. In such circumstances, development of new watering-points in areas which could scarcely be used before due to lack of water, can only lead to the overgrazing and deterioration of those areas. Where previously they could have supplied some grazing during the wet season without too much damage, this possibility is ruled out if overgrazing is facilitated during the dry season. As many instances have shown, development of water supplies in these areas can at best provide temporary relief until livestock numbers catch up with the capacity of the newly available range. The end result, however, is that problems once confined to part of the rangeland in question will have been spread over the whole of it.

Water development as a means of improving forage conditions for wild game in national parks is equally risky. If animal numbers and distribution cannot be controlled, any area opened up is likely to become overgrazed and deteriorate.

If the full cooperation of those who control numbers and distribution of animals can be obtained, however, water development is an important means of increasing the productive capacity of the land. Particularly useful is the provision of temporary watering-points, able to be closed down with the first signs of overuse and range·deterioration. A series of such points opened in rotation can be used to shift animal concentrations from one area to another and achieve more uniform use of the rangeland. However, without the cooperation of the graziers, it is virtually impossible to close a watering-point once it has been made available, so that the same sort of destruction may result as from a development project that was intended to be permanent.

(e) Use of Chemicals

A great variety of chemical herbicides are now available for the control of undesirable vegetation. Some of these are specifically for the control of a particular kind of plant, others of the broad-spectrum variety killing, for example, all broad-leaved plants. Some must be applied directly to the plant, others can be sprayed more indiscriminately. The obvious advantages offered by such chemicals in pasture management has encouraged their widespread use before their full environmental effects could be studied. As noted elsewhere (Chapter 6), there have been indications that one of them, 2, 4, 5-T, may produce harmful effects on animal life, including man. Others have not yet been fully tested.

All that can be said at this time is that herbicides which affect a wide variety of plants and are persistent in the environment, should be used with the greatest care, and fully tested in a trial area of the local vegetation before being used extensively. If this is not done, the consequences may include destruction of desirable plants in the area treated or in areas to which the chemical accidentally spreads; destruction of aquatic vegetation in water bodies to which the chemical is carried by run-off; and possible adverse effects on animal populations and on man.

(f) Planting and Reseeding

As already mentioned (section 2(f) p. 86), natural succession will bring about recovery in range condition that has not been allowed to deteriorate beyond a certain threshold, once the causes of deterioration are removed. If the threshold has been passed, however, the natural processes of recovery may operate far too slowly to be useful in management. Reseeeding or replanting of rangelands is then the most effective way to improve conditions. A great variety of rangeland plants have been developed by plant geneticists and breeders, and these are suited to a wide range of climatic and soil conditions. Selection of the group of species most likely to produce good results in a particular area requires a high degree of expertise. Trial and error can be both expensive and risky.

Two general principles need to be stressed.

1. Reseeding will be a waste of time and money unless grazing of the area can be controlled. While the new range plants are becoming established, they will usually require nearly complete protection. Thereafter, grazing must be kept within the limits of seasonal and annual carrying capacity.

2. A badly deteriorated range usually will not support the range plants that ultimately will be desired in order to provide maximum forage yield. It is therefore often desirable to start the process of recovery by reseeding with low successional species, even so-called weeds, which will help to build up soil cover, prevent erosion and improve the site to the point where better quality range plants can be successfully established.

(g) Control of Animal Diseases

Some areas of potential livestock range are at present not used by livestock because of the prevalence of animal disease for which treatment or control methods have not been available. In extensive areas used by grazing animals their productivity is kept low by maladies affecting their general health and reproduction. It is apparent, therefore, that elimination or control of disease is a means by which both the area available to livestock and the potential production from existing livestock ranges may be increased.

Within this context, much attention has been paid to the elimination of the tsetse fly in Africa and the opening-up for livestock of potential grazing areas occupied by the fly. It is, of course, the vector of the trypanosome which causes the disease known by its Zulu name *nagana*. The fly usually requires woodland or scrub as shelter, and disappears when woody vegetation is removed. It feeds and depends for survival on wild game when domestic livestock are not present, and most wild species are resistant to, or not much affected by, the trypanosome. Methods for control of tsetse fly have involved both the shooting-out of wild game and the clearing of scrub and woodland, or sometimes both. Pesticides have also been widely used in attempting to control tsetse fly.

These control methods have encountered the following difficulties. (1) In areas of woodland or scrub it is extremely difficult to shoot-out all the wild species that can serve as hosts for the fly. Elimination of the conspicuous or vulnerable animals alone will not eliminate the fly. Furthermore, unless a species is exterminated, it will often recover quickly from the effect of a shooting campaign, unless its habitat has been destroyed. Shooting, therefore, becomes an endless process with uncertain results. (2) Destruction of woodland and scrub can have the effect of destroying the habitat both for tsetse and for the more hunting-resistant species of game. If the area is opened up, game can more readily be kept out by shooting. However, normal successional processes commonly lead to reinvasion by woodland and scrub, unless the area is brought under intensive management and constant attention is given to preventing such re-invasion. (3) Destruction of game animals and of their habitat is destruction of a valuable resource. Any project for replacement of game by livestock should be preceded by the most careful assessment of the comparative values over the long run of the two kinds of animal. It may well be that the game could produce more return from the land over the longest period of time. This is almost certain to be true if the alternative is unmanaged and uncontrolled use of domestic livestock, since in these circumstances the land will be overgrazed and its long-range productivity destroyed. (4) Use of persistent pesticides to destroy tsetse flies brings in all of the adverse environmental effects of these substances discussed later in Chapter 6. Furthermore, there is no guarantee of their long-term effectiveness.

The same basic set of problems affects, in varying degree, the opening-up of any area previously unusable by domestic animals because of the prevalence of disease. Furthermore, control of diseases which occur in areas actually used by livestock and which tend to diminish the health or productivity of the animals, may remove the only barrier to their rapid increase in numbers to the point where the areas will be overstocked and overgrazed, and seriously deteriorate.

The answer to these problems is not to give up the control of disease but, rather, always to accompany the control of disease with the institution of sound land and animal management programmes. Methods of disease control should be used with full cognizance of their broader environmental effects. In no case should a valuable wild resource be destroyed to permit its replacement with a less valuable domestic resource. The value of disease control is likely to be appreciable only on the better quality rangelands where intensive management of both vegetation and animals is both ecologically feasible and economically profitable.

The principles involved are once again conveniently summarized in the paper presented to UNESCO's Man and the Biosphere Conference in 1968, to which previous reference has been made.

'The introduction of improved exotic stock to areas in which diseases have not been controlled is a move that is certain to fail.

'The curing of animal disease is usually not an end in itself. Unless simultaneously accompanied by improvements in feeding, management, and breeding, the advantages of an improved health situation can be lost. There is no point in preventing animals from dying of disease if they are then to die of starvation' (Tribe *et al.*, 1970).

(5) ALTERNATIVE APPROACHES TO RANGELAND DEVELOPMENT

The productivity of any rangeland will vary with the quality of the site and with range condition. The best results can be expected if development efforts are concentrated on the best sites, with the deepest, most fertile and productive soils, and with the most reliable climates. Intensive management of small areas of highly productive rangeland will give better returns than extensive management of much larger areas. Thus development of irrigated pastures or intensively managed permanent pastures where soils and climate are favourable will be more rewarding than the expenditure of greater effort on large scale projects.

Greatly increased yields may be obtained through improvement of range condition. A change from very poor to excellent condition, through proper management, may increase carrying capacity sevenfold—in other words, one acre restored to excellent condition will support as many animals as seven acres left in poor condition. These figures are based on average quality annual grass ranges and can be exceeded on higher quality perennial grass ranges.

It makes much more sense, economically and ecologically, to invest development money in improving the quality of existing rangelands by reversing downward trends and upgrading range condition, than it does to bring new areas of land into livestock production. This is particularly true where new lands would merely be subjected to the same systems of management that caused depletion on already established areas.

In any project for rangeland improvement particular attention needs to be paid to wild land and wildlife values. These can be exceedingly high. A development plan that concentrates livestock production on the most suitable areas and keeps other areas for wildlife may give a far better return than one which involves the less intensive development of an entire region without reference to the preservation of its full potential of uses and values (Figure 13).

East African experience may be used as an example to illustrate these points. The principles for effective rangeland development in that region have been conveniently summarized by Heady (1960), in the following quotations, of which the first has been largely covered in the foregoing discussion, but the other two underline additional points which it is important to keep in mind.

'The first and by far the most important step is to gain control of livestock numbers. Excessive grazing pressure has caused vast deterioration of the soil by erosion. Improvement cannot be attained until grazing is made

Figure 13. With minimum available water species such as the scimitar-horned oryx can produce protein from the desert edge. (Photograph by F. Edmond-Blanc).

light enough to allow the soil to once again become covered with palatable grasses. Without grazing control none of the other range management practices can give more than temporary relief. Experience has shown that money put into schemes without the grazing being done properly will be lost, and, frequently, worse conditions have developed than originally existed.

'Along with grazing control there must be an adequate marketing structure to handle the animals removed in a destocking programme and in normal production afterwards. Destocking should be done on a rotational basis, otherwise markets have no chance of handling the number of animals that need to be removed. On this basis each district should be appraised and the number of animals reduced accordingly.

'(Another) aspect of rangeland development concerns the integration of stock raising and development of land for crops. Opening rangeland for cultivation is doomed to failure because the climate is such that crops will not grow every year. . . . The limiting factors are soil and climate, not what people need. It is usually better to use money to improve production from existing cultivated land than to open rangeland for cultivation.'

Riney (1972), in considering criteria for land-use planning, has emphasized the need for cross-checks to determine whether a recommended development is suited to an area under consideration. The essential principles are stated in the following quotations.

'An ecological test is given as an example; the test is simply to determine to what extent present forms of land use are meeting the minimum requirements for conservation. This "conservation criterion" as it applies to pastoral lands assumes that unless some sort of stable vegetative cover can be maintained there can be no sustained production. In pastoral terms, if the habitat elements on which the wild or domestic species depend are downgrading, then the animals are too numerous. The area is overstocked; the present pattern of land use is unsatisfactory.

'Socially and ecologically, rapid large-scale changes are best avoided, and gradual development is much to be preferred. . . . Gradual development inevitably brings into play various intercompensatory mechanisms, or homeostatic mechanisms, which in turn produce evidence useful in evaluating trends. Thus, if domestic or wild animals are overstocked, the imbalance between plants and animals may be detected in vegetation precisely by recording changes in species composition, or in grosser terms, by noting the changes in the proportions of perennials, annuals, shrubs and trees. Consequently, animal responses involve, for example, loss of physical condition and increased mortality of young. While changes are slow, these kinds of evidence can quickly and simply be recorded in time to recognize and reverse the trend. However, if high numbers of animals suddenly occupy land of low carrying capacity, changes can be so rapid that, by the time evidence of the trend is clearly appreciated by land managers, consequences may be catastrophic and irretrievable.'

Obviously, any plan for rangeland development needs to be preceded by a thorough survey of the resources and land use capabilities of the region to be developed. Such surveys have commonly been employed, but at least two elements have virtually always been lacking: first, the participation of experts capable of evaluating the existing wild land resources of the region —vegetation, animal life, scenic and recreational resources—from the viewpoint of their potential contribution to economic development; and, secondly, the contribution of ecologists trained to assess the region as a functioning ecosystem or group of ecosystems, both in its wild or undeveloped state and in the various derived states that would result from proposed developments. Usually, also, survey teams lack expertise in anthropology, sociology and other sciences concerned with human cultures and behaviour. In consequence, wild land values are too frequently sacrificed for doubtful gains that involve the elimination of wild species and their replacement with domestic forms. Another frequent result is that the broad environmental consequences of the proposed developments are ignored, while the receptivity of the people in the area and their probable responses to the proposed developments are insufficiently considered.

Undeveloped wild rangelands have values not only in themselves but also in the contributions they can make to the better management and utilization of those areas that are to be modified and changed for human use and occupancy. In the former category are the scientific, educational and aesthetic or recreational values, while the derived values include not only the potential of their genetic resources, but also the knowledge of ecology that can be applied to better management of the areas to be modified.

Since wild areas are becoming increasingly scarce, their value increases proportionately with passage of time and the disappearance of comparable areas (Figure 14). There is only one appropriate time to establish national parks and reserves and that is before land use changes are accelerated

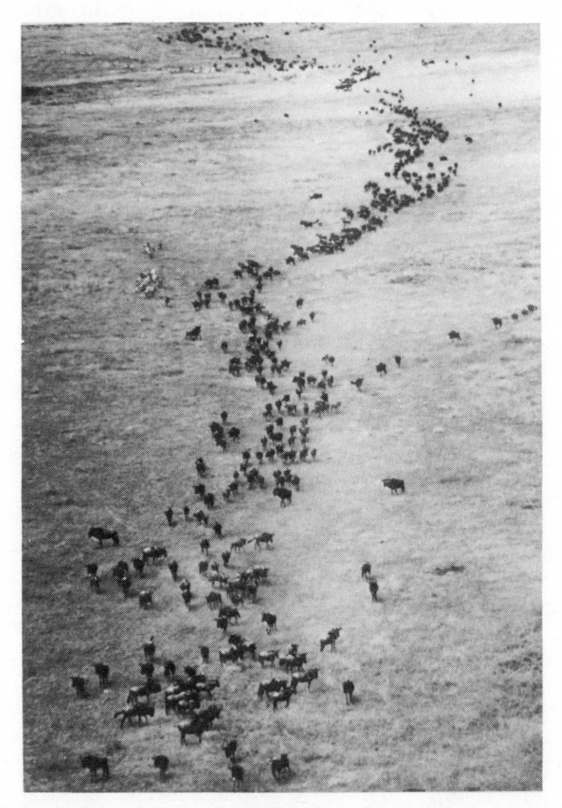

Figure 14. Since wild areas are becoming increasingly scarce, their value increases proportionately with passage of time and the disappearance of comparable areas. Wildebeest in the Serengeti. (Photograph by Norman Myers).

by development, because it is exceedingly difficult to restore the full range of values to areas that have been modified in the development process.

It is essential that such reserves be established not only in the lands that have the least value for the economic production of domestic animals or plants, but also in those that are most valuable for that purpose, since they will contain species, communities and ecosystems that will be absent from poorer areas. Admittedly, in such high-value areas natural reserves are bound to be relatively small in comparison with the lands to be otherwise developed, but the contribution they can make to a proper understanding of how the development areas should be used and managed will be equivalently high.

For these and other reasons to be discussed in the next chapter, it is unfortunate that inadequate attention has yet been paid in most developing regions to the high potential economic value of national parks and similar areas both directly through tourism and recreational use and also for their other values. As a development alternative in the land use planning of rangelands it deserves the most careful consideration.

(6) REFERENCES

Albrecht, W. A. (1947). Soil fertility and biotic geography. *Geographical Review,* **47(1),** 86–105.

Albertson, F. W., Tomanek, G. W., and Riegel, A. (1957). Ecology of drought cycles and grazing intensity of grasslands of the Central Great Plains. *Ecological Monographs,* **27(1)**, 27–44.

Colman, E. A. (1953). *Vegetation and Watershed Management.* Ronald Press, New York.

Cooper, C. F. (1961). The ecology of fire. *Scientific American* **204(4),** 150–160.

Dasmann, R. F. (1964). *African Game Ranching.* Pergamon, Oxford.

Dasmann, R. F. (1964). *Wildlife Biology.* John Wiley, New York.

Dasmann, R. F., and Mossman, A. S. (1962). Abundance and population structure of wild ungulates in some areas in Southern Rhodesia. *Journal of Mammalogy,* **43,** 533–7.

Dasmann, William P. (1945). A method for estimating carrying capacity of rangelands. *J. Forestry,* **43(6),** 400–2.

Dasmann, William P. (1948). A critical review of range survey methods and their application to deer range management. *California Fish and Game,* **34(4),** 189–207.

Dasmann, William P. (1951). Some deer range survey methods. *California Fish and Game,* **37(1),** 43–52.

Draz, Omar (1954). Range management—a world problem. *J. Range Management,* **7(3),** 112–14.

Dyksterhuis, E. J. (1949). Condition and management of rangeland based on quantitative ecology. *J. Range Management,* **2,** 104–15.

Elton, Charles (1959). *The Ecology of Invasions by Animals and Plants.* Methuen, London.

Giles, Robert H., Jr., ed. (1969). *Wildlife Management Techniques.* The Wildlife Society, Washington.
Graham, Edward H. (1944). *Natural Principles of Land Use.* Oxford, New York.
Heady, Harold F. (1960). *Range Management in East Africa.* Kenya Department of Agriculture and EAAFRO, Nairobi.
Humphrey, Robert R. (1962). *Range Ecology.* Ronald Press, New York.
Koford, Carl (1958). Prairie dogs, white faces and blue grama. *Wildlife Monographs,* No. 3.
Linsdale, Jean M. (1946). *The California Ground Squirrel.* University of California Press, Berkeley.
Odum, Eugene P. (1959). *Fundamentals of Ecology.* W. B. Saunders, Philadelphia.
President's Science Advisory Committee (PSAC) (1967). *The World Food Problem,* Vol. I–III. White House, Washington.
Puri, G. S. (1960). *Indian Forest Ecology.* Oxford, New Delhi, two volumes.
Ratcliffe, Francis (1959). The rabbit in Australia. In *Biogeography and Ecology in Australia. Monog. Biol.,* **8,** 545–64.
Reynolds, Hudson (1956). The ecology of the merriam kangaroo rat (*Dipodomys merriami* Mearns) on the grazing land of southern Arizona. *Ecological Monographs,* **28(2),** 111–27.
Riney, Thane (1955). Evaluating condition of free-ranging red deer (*Cervus elaphus*), with special reference to New Zealand. *New Zealand J. Science and Technology.* **B36(5),** 429–63.
Riney, Thane (1956). A zoo-ecological approach to the study of ecosystems that include tussock grassland and browsing and grazing animals. *New Zealand J. Science and Technology,* **37(4),** 455–72.
Riney, Thane (1972). An ecological approach to aid to developing countries. In *The Careless Technology.* Natural History Press, New York.
Rudd, Robert (1964). *Pesticides and the Living Landscapes.* University of Wisconsin, Madison.
Sampson, A. W. (1952). *Range Management Principles and Practice.* John Wiley, New York.
Savory, C. A. R. (1969). Crisis in Rhodesia. *Oryx,* **10(1),** 25–30.
Stamp, L. Dudley, ed. (1961). *A History of Land Use in Arid Regions,* UNESCO, Paris.
Stoddart, L. A., and Smith, A. D. (1955). *Range Management.* McGraw-Hill, New York.
Strahler, Arthur, N. (1970). *Introduction to Physical Geography.* 2nd Edition, John Wiley, New York.
Talbot, Lee M. (1964). The biological productivity of the tropical savanna ecosystem. In *The Ecology of Man in the Tropical Environment.* IUCN, Morges.
Thomas, William L., Jr., ed. (1956). *Man's Role in Changing the Face of the Earth.* University of Chicago, Chicago.
Tribe, D. *et al.* (1970). Animal ecology, animal husbandry and effective wildlife management. In *Use and Conservation of the Biosphere.* Proceedings of the Intergovernmental Conference of Experts on the Scientific Basis for Rational Use and Conservation of the Resources of the Biosphere, Paris, 4–13 September 1968. Natural Resources Research X, UNESCO.

UNESCO/FAO (1968). *Conservation and Rational Use of the Environment.* Report to Economic and Social Council of the United Nations, ECOSOC, 44th Session, Doc. E/4458.

UNESCO (1970). *Use and Conservation of the Biosphere.* Natural Resources Research X, UNESCO, Paris.

Weaver, John E., and Clements, F. E. (1938). *Plant Ecology.* 2nd Edition, McGraw-Hill, New York.

Wellington, J. (1955). *Southern Africa: A Geographical Study.* Cambridge, New York, 2 volumes.

CHAPTER 5

Development of Tourism

(1) INTRODUCTION

Tourism is receiving increasing attention from international development agencies and national governments. It can generate much needed foreign exchange for financing other sectors of a developing economy; investment in tourist facilities can be financially profitable, with high rates of return. In OECD member countries alone, tourist arrivals in 1969 were about 120 million and foreign currency receipts totalled $12·3 billion (OECD, 1970). Mexico has projected tourist expenditures of U.S. $1·6 billion for 1980 (*Focus on Mexico*, February 1970), and investment in tourism in Mexico is claimed to create about 75 per cent more jobs than equal amounts invested in petroleum, metals, or electronics industries (*Los Angeles Times*, 19

April, 1970). Tourism investment may, however, have to include public expenditure for infrastructure, which may result in an economic cost-benefit ratio less favourable than that suggested by financial returns from specific projects such as hotels.

Tourism development poses special ecological problems not encountered in other types of economic activity. The environmental resources 'exploited' for tourism attract visitors because of their outstanding beauty, recreational possibilities or educational interest. Often, as in high mountains and islands, the resources of interest for tourism are readily damaged by disturbances (Figure 15).

Where the focus is on natural areas with exceptional wildlife, such as the Galapagos islands, the economic goals of tourism must be adjusted to the ecological and biological dictates for managing these areas. The urgent need to establish reserves for scientific purposes is, in fact, closely and positively linked to the proper management of natural areas for tourism. National Parks can provide an effective buffer between surrounding settled regions and biological resources located within Park confines.

The environmental amenities which attract tourists have tended to be taken for granted, and the preservation of their quality has only recently

Figure 15. Tropical shore in Ceylon. Such areas have great appeal for tourism, yet their development must not destroy the scenic or biotic values that originally cause tourists to visit them. Careful planning is vital if development and conservation are to move forward together. (Photograph by F. Vollmar: courtesy World Wildlife Fund).

begun to concern tourism development planners. A certain amount of deterioration in environmental quality may be viewed as a necessary trade-off in some kinds of economic activity, such as mining and manufacturing. However, for tourism the quality of the environment is the basis for attracting visitors and must be conserved.

(2) REGIONAL PLANNING CONSIDERATIONS

The potential of a region for tourism may be so great that its development will be central to economic goals and plans. However, regional planning for tourism more often will be considered as a separate and distinct planning effort, aimed primarily at determining the best location of hotels, resort facilities and transportation networks. Such planning views tourism primarily as a source of foreign exchange. This approach would recommend development to accommodate the maximum possible number of projected visitors and to design facilities to generate the maximum possible spending. However, a profit-maximizing orientation to tourism development can result in the deterioration of fragile ecosystems and attractive landscapes through overbuilding and excessive densities of visitors. Also, the concentration of tourists in a small number of large hotels owned and managed by outside entities is liable to result in 'leakage' of earnings out of a region or a country. The aims of conserving the environmental amenities of a region and of advancing regional development through tourism are interdependent. The more local people benefit from tourism, the more they will benefit from a commitment to preserve the environmental features which attract tourism. Consideration of the region's ecological characteristics are essential to provide adequate criteria for the design of facilities and viable plans for the use and management of resources for tourism.

The disadvantages of unplanned or spontaneous development of tourism are well illustrated by Yugoslavia's Adriatic coast (Gasparovic, F. *et al.*, 1971). Despite a measure of governmental control during its development the great increase in numbers of visitors to the coast resulted in rapid, unplanned construction of hotels and dwellings both for tourists and for migratory service employees. This expansion threatened to destroy the very environment that is the tourist attraction and stimulated the preparation of a long-range development plan for tourism. This example underlines the environmental threat, both aesthetic and biological, from overcrowding of particularly attractive areas. Physical planning can help to avert such undesirable consequences.

There is a carrying capacity for tourists, as well as for any other type of use, which will vary with the fragility of the area concerned and the nature of the tourist activity contemplated. Thus, an island rainforest is a fragile environment which could be endangered by excessive numbers of people; enjoyment of such an environment by visitors necessitates a degree of solitude. By contrast a recreational beach is not easily damaged by sunbathers or swimmers and many people are happy with a certain degree of crowding in such areas. Planning facilities for tourism in the previous case would require limitations on hotel, motel and similar installations in the region concerned, in order not to encourage excessive numbers of visitors, whereas in the latter instance high density accommodations for tourism might well be tolerable.

(a) Evaluating Resources and Landscapes Suitable to Tourism

This process is closely related to integrated natural resource surveys and evaluation studies, the methodology of which has evolved over the past century. A focus on tourism resource use introduces special value considerations, which are usually established subjectively; nevertheless, the relevant cultural values can often be identified and quantified. Preferences of various segments of a population can be surveyed and patterns of recreation can be observed and related to demographic data. The active expression of preferences, such as swimming, hiking, camping and sight-seeing, can be assumed to be both a function and a reflection of socio-economic characteristics. These preferences are ultimately expressed as demand on different types of recreational resources or landscapes, which in turn influences the classification and evaluation of the various resources in a region.

International tourism, by definition, draws persons from diverse cultures; regional resource inventories and evaluations for the development of tourism should therefore attempt to take into account diverse recreational preferences. The travel preferences of tourists from different cultures (or groups within cultures) should also be considered in the layout and design of tourist facilities. Particular attention is needed to cultural preferences for the total environment destined for tourism use. Such considerations will again assist in assigning specific values to different landscapes and natural resources.

A questionnaire survey of tourists in Jamaica revealed two basic and possibly conflicting attitudes (Anon, 1960). One was a desire to be surrounded by a different and romantic culture; the other was a desire to have all the comforts of home, such as air conditioning, good transport and well-stocked drugstores. Another finding was that winter tourists with generally higher incomes were content with the 'American Plan', while the less rich summer tourists were more intent on sightseeing and preferred more modest and flexible hotel arrangements.

The more active tourists (e.g. hikers, swimmers, boating enthusiasts, campers) generate a special demand on resources and landscapes that should be given proper weight in the regional inventory and evaluation.

The elderly, affluent tourist often creates the smallest environmental demand, yet spends the most. Such preferences should not be viewed as axiomatic, however, since tastes change and there is a definite trend in demand towards naturalness of surroundings, both cultural and physical, as international travel costs decrease proportionately to discretionary income, as greater numbers of young people travel, and as understanding of other cultures increases.

Particular attention should be paid to the recent increase in scientific and educational tourism, in which the people concerned seek opportunities for nature study and the development of understanding of unique or at least different kinds of ecosystems and assemblages of wild species. Such tourists can be accommodated in areas from which mass tourism might best be excluded.

On a regional level, it is the preferences of the more active outdoor recreation-oriented tourists which are important for classifying and evaluating resources, and which lend themselves to an ecological approach in survey work. Environmental tourism planning takes into account relationships of different aspects of physical environment. An investigation undertaken in the state of Wisconsin, USA, is illustrative (Lewis, 1969, 1966). Environmental resources were inventoried to identify recreational potentials. Environmental features, such as wildlife complexes, vegetation, streams and lakes, waterfalls, caves and historic sites, were found to occur in distinct patterns, termed environmental corridors, whose combined features had more recreational significance than any individual resource pattern. Within the corridors, more detailed ecological studies were carried out in order to establish recreational priorities. Recommendations for environmental preservation and management of important recreational complexes were based on the findings and the priorities assessed accordingly.

The ecological impact of development activities on the coastal zone requires specific attention in tourism planning. One imaginative approach to estimating the possible impact of different activities on coastal environments has been developed by the state of California for its Comprehensive Ocean Plan (COAP, 1970). Cause and effect relationships associated with different land uses (and the specific activities under each land use) were considered in minute detail, then presented in chart form. These charts relate specific actions to possible adverse environmental impacts, which are in turn classified according to initial condition, consequent condition and final effect.

(3) THE DEVELOPMENT OF NATIONAL PARKS

The national parks of East Africa exemplify the great potential that parks can have for attracting international tourists, and thereby for contributing to the foreign exchange earnings of a country. Hart (1966) and others, however, have cautioned against overemphasizing tourism as a factor in the justification of national parks, since the requirements for protection of the flora and fauna of such areas may exclude the possibility of accommodating large numbers of visitors.

(a) General Concepts

National parks represent a particular kind of land use intended to permit the maximum appreciation of protected areas of high value because of the nature and quality of their flora, fauna or landscapes.

A national park is defined as a relatively large area (1) where one or several ecosystems are not materially altered by human exploitation and occupation, where plant and animal species, geomorphological sites and habitats are of special scientific, educative and recreative interest, or constitute a natural landscape of great beauty; (2) where the highest competent authority of the country has taken steps to prevent, or eliminate as soon as possible, all exploitation or occupation and to enforce effectively a proper respect for the ecological, geomorphological or aesthetic features which have led to its establishment; and (3) where visitors are allowed to enter, under special conditions, for inspirational, educative, cultural and recreative purposes.

Exploitation of natural resources, including agricultural and pastoral activities, hunting and fishing, lumbering, mining, and dam construction for the purpose of irrigation or hydroelectric power, are normally excluded from national parks. Residential, commercial or industrial occupation, and the building of roads, railroads, aerodromes, ports, power lines, telephone lines, etc., are also generally excluded from a national park. Exceptions to these restrictions are limited facilities to permit effective park administration and management, and to permit access and use for appropriate outdoor recreation by park visitors.

Tourism and recreation from hiking, swimming, boating, mountain climbing, wildlife viewing, photography and other non-consumptive uses are encouraged in national parks, except within restricted areas set aside primarily for species and ecosystem protection and to be used by qualified scientists for approved research purposes. Certain types of outdoor recreation, involving construction of elaborate facilities, such as ski-lifts and jumps, or the use of motorized transport—motor boats, dune-buggies, snowmobiles, etc.—are generally unsuited to national parks.

Mass use areas for recreation and tourism should, whenever possible, be located outside of national parks or equivalent reserves. These include hotels, motels, resorts and associated facilities, and organized recreation grounds (for tennis, golf, etc.).

Construction of roads, telephone lines and other transportation or communication facilities, needs to be kept to the minimum required for effective management or for a reasonable degree of visitor use of the park, bearing in mind the primary purpose of protection and management of its resources. Even buildings and other headquarter facilities for park personnel should, whenever feasible, be located outside the park boundaries.

Although hunting and fishing should normally be excluded from a national park, special exceptions may be made where removal of animals is essential for the maintenance of balanced biotic communities or for the overall protection of park values. But such hunting and fishing would be permitted only under strict control by the designated park authorities. A further exception may be made by allowing recreational fishing in certain designated park areas, provided that adequate attention is also given to the maintenance of unexploited aquatic habitats within the park.

Cutting, burning or other removal of vegetation within a national park should only be permitted when scientific study has shown it to be an essential management tool in the maintenance of biotic communities that would otherwise disappear. Existing private ownership and special rights (e.g. grazing) within a national park may be accepted so long as they are confined to a small area and their redemption or termination is foreseen within a reasonable period of time.

Any park of appreciable size should have within its boundaries:

1. One or more strict nature reserves, from which the public is excluded;
2. Extensive wilderness areas within which there are no developments and which are accessible to the public only by primitive means of transport—hiking, horseback, canoe or other motorless boat, etc.;
3. Areas set aside for mass tourism, which the general public can reach by public transport or private motor vehicles, or which are close to roads and easily accessible by hiking trails; and
4. Intensive use areas, such as camp grounds or similar tourist facilities, where these cannot reasonably be located outside the park boundaries.

The ratio between these categories will be dependent upon the nature of the biotic communities to be preserved, but normally the first two could be expected to occupy the larger part of the park and the other two only a relatively small area.

In surrounding areas protection may be needed almost to the same degree as within national parks themeslves. Thus a park located in the middle of intensive land use or human settlement is most difficult to maintain. Conflicts between internal and external values may become extreme, particularly where animal movements from the park are concerned. Where such a situation is unavoidable, expensive fencing or the construction of other barriers will usually prove essential.

Parks are most viable when buffered by lands that are used only for extensive rather than the more intensive forms of exploitation, such as forestry, ranching or hunting and other kinds of public recreation.

Where large wild animal populations are involved there is value in having a controlled hunting zone immediately around the park in which the populations of these animals can be regulated. It will also serve to minimize conflicts with land use in the areas beyond.

National parks are not intended, and should not be used, for all mass outdoor recreational needs. Ideally, they should constitute one end of a spectrum of recreational areas, the opposite end of which will be represented by city parks and playgrounds. With proper attention to the development of suitable mass recreational facilities in areas outside their boundaries, national parks will be subjected to recreational pressure only from those who seek the natural values that parks can provide.

Park administration and management are technical skills, requiring highly qualified personnel. National park work at all levels requires a special quality of interest and devotion from those employed in it. If the supply of park personnel of this calibre is to be assured, schools and universities equipped to train them are essential and, ideally, ought to be available in every country. However, as a minimum requirement, the establishment of regional training centres serving a group of countries with similar conditions will suffice.

Every park has a carrying capacity for visitors dependent upon the ability of its landscape, flora, and fauna to withstand visitor use without deterioration. This sets the limits on development of roads, trails, camp grounds, hotels, motels or other tourist facilities. Visitor use, in excess of a park's capacity, should be discouraged by every possible means, both to protect the resources of the park and also to maintain and to retain the quality of outdoor recreation at a high level.

(b) Park Planning

Integrated surveys for planning natural resource development offer an excellent opportunity to identify potential national park areas, as well as other areas whose uniqueness justifies preservation. These surveys are usually the first step in the economic development of regions whose resources are little known, although not necessarily unexploited.

It is always important to relate park planning to total recreational planning, in order to provide for varied recreational uses and to avoid concentrating people within national parks, where overcrowding would have adverse effects. The planning should be related, in turn, to the overall economic plans for the region concerned. This implies that knowledge of recreational patterns should be translated into resource classification and evaluation. Features of cultural or historical importance, such as archaeological sites, monuments and important battle grounds, should of course be included in the inventory to complete the picture.

In any systematic park planning at a national level, therefore, the following steps are recommended (after Hart, 1966):

1. a reconnaissance survey to identify high priority sites and projects;
2. beginning with these high priority sites, a study of individual recreation resource sites to obtain a more detailed knowledge of them;
3. collaboration with regional planning authorities to assess relative values of different combinations of land use, and to incorporate park plans into regional economic plans;
4. in the light of the decisions reached on the park system to be established, preparation of specific development plans for each area, covering research needs, management of resources and visitors, interpretive programmes, etc.; and
5. recruitment, training and all the necessary provisions for the support of the personnel required to manage the park system.

(c) Park Management

Data from basic and applied research on the flora, fauna and total physical environment and ecology of parks are essential for elaborating a sound management policy..The latter will cover such questions as the degree of purposeful intervention in the habitats and wildlife of a park, the fixing of boundaries in relation to ecological requirements of wildlife, the level and kind of permissible recreational activities, the location of areas to be made accessible to tourists, the siting and capacity of tourist facilities, the design of structures and roads, and the educational and interpretive programmes to be provided (Figure 16).

Policy decisions on these and other questions should be incorporated in management programmes which in turn will establish criteria for the number and qualifications of park personnel, construction of roads and other publicly operated facilities and the scale of initial and on-going park research. The development and funding of this infrastructure are particularly important if a park is intended to be a major focus for tourism.

Experience in East African parks indicates that management of park habitats and wildlife should be kept at an absolute minimum, and that management activities such as burning vegetation or shooting wild animals are justifiable only when irreversible damage is otherwise likely (Russell, 1970). The impact of visitors must also be considered. For example, in the Serengeti National Park, Tanzania, with an area of about 5000 square miles, seven more lodges are planned to accommodate park visitors. They

Figure 16. The combination of natural terrain and vegetation and the cultural interest of an old Incan city provide a justification for careful management of the Machhu Pichu site in Peru. Excessive development for tourism could destroy those values. (Photograph by H. Jungius: courtesy World Wildlife Fund).

would be located not less than 10 miles from the park boundaries and have not more than 150 beds each. A limit of four people per safari car is contemplated and not more than 200 vehicles would be permitted to operate for all park safaris together. These limits are designed to keep the visitors at a level which will not overwhelm the habitat and the quality of the park experience. They are, however, at this stage, the result of judgements rather than precise determinations of carrying capacity for park visitation.

The relevant considerations have been succinctly stated by Houston (1971). 'The maintenance of a natural park ecosystem requires a unique approach to research and management. Unlike other forms of land management, management of a park ecosystem generally involves preventing or compensating for man's altering of natural ecological relations. . . . Providing for the educational and aesthetic enjoyment of man, while maintaining pristine ecological relationships, represents the greatest challenge in the management of natural areas.'

Some of the pitfalls and dilemmas inherent in the pursuit of this policy in the US national parks are explored by Darling and Eichhorn (1969). They include excessively high densities of park visitors, destruction of fragile sites traversed by trails, with special reference to the propensity of visitors for picking flowers and collecting rock specimens in alpine zones, unaesthetic design and siting of park buildings, and the disadvantages of having accommodations within a park as opposed to outside its boundaries. In countries with limited numbers of adequately trained park personnel, the major management concerns are likely to be safeguarding the fauna and flora of the ecosystem from local poaching, excluding squatters and, only after these objectives are achieved, the reconciliation of ecosystem integrity with recreational uses.

However, the eventual enjoyment of parks by domestic and international visitors must be contemplated and planned for, and in general the sooner that active programmes are staffed and budgeted, the sooner will the parks be so enjoyed. Moreover, the development of tourism should certainly serve to generate opportunities for employment as guides and park staff and in various ancillary services for quite a number of local people who might otherwise engage in illegal hunting, timber cutting, cultivation and other disruptive uses of the park ecosystems. This indirectly favours other goals of park management, such as the conservation of these ecosystems and provision for scientific research.

Among these other objectives reference should be made to the function of some parks as biological preserves, where gene pools of flora and fauna may be maintained. In parks or sectors of parks suited to this purpose, a more restrictive management policy is indicated, with special reference to the regulation of plant collecting, hunting and, perhaps, fishing. In order that an appropriate management and development policy may be worked out, it is important that the value of a particular area for the preservation of certain biological resources should be established at the outset of planning. The significance of gene pools, which can hardly be overestimated, is discussed in Chapter 6, in relation to the disappearance of genetic resources important to agriculture. It need only be emphasized here that similar considerations apply to the wild progenitors of domesticated animals and to other wild species whose future value to man is unknown. Thus, wild ungulates are recognized as an existing or potential source of animal protein (Talbot, 1966), and could well have genetic characteristics which may prove to be valuable. The same is true of most plant species in tropical rain forests, the potential worth of which is still often unknown. The fact that numbers of tropical plant species have already been found to be of inestimable pharmaceutical value is a good indication of the benefits that can follow the perception of the usefulness of a particular plant. As Stone (1969) remarked, the genetic diversity in plants inherited by modern man is the result of hundreds of millions of years of evolution, so that the loss of any species represents the loss of information, and reduces the richness of the community of all life.

(d) Parks and Tourism in Tropical Mountains

Tropical mountains of high elevations contain a wide range of environments, and may have landscapes of unexcelled beauty as well as of great scientific interest (Figure 17). The potentials for park-oriented tourism development—based on both domestic and international visitors—are correspondingly considerable. In the tropics, mountain resorts and lodges are often the only respite from the heat and humidity at sea level (Withington, 1961). In addition, elevations above the tree line (which is sometimes rather ill defined on tropical mountains) contain the rugged terrain and snow-covered peaks which attract hikers and climbers.

Two general considerations are important to assessing the tourism potential of tropical mountains: altitude and the nature of the rainfall. In tropical latitudes, temperatures characteristic of montane and alpine zones

Figure 17. Wild vicuna and unique Andean highland terrain combine to give scientific and touristic values to the Pampa Galeras, Peru, that exceed the value of any other form of land use. (Photograph by William Franklin and A. Stokes).

occur at much higher elevations than elsewhere. Thus, although the peaks may have perennial snow, their recreational potential for 'winter sports' is very low, since they will probably be at least 4,000 to 5,000 metres above sea level. For most people, especially the elderly, difficulty in breathing accompanied by fatigue may be experienced on first arrival at elevations above approximately 3,000 metres and several days may be needed to acclimatize. This is a limiting factor on the location of hotels and resorts. Rainfall patterns tend to exercise further limitations. Maximum rainfall in mountains rising from tropical lowlands generally occurs between 1,500 and 2,000 metres above sea level, and consequently the zone of ideal climatic conditions for tourist resorts tends to be the rather narrow one from 2,000 to 3,000 metres. This may be compensated by superb conditions, in particular on the leeward of a tropical massif, which may be relatively dry, as in the case of the Sierra Nevada de Santa Marta on the Caribbean coast of Colombia.

The maintenance of parks in alpine zones demands that careful attention be paid to the fragility of the ecosystems involved. Plant growth is largely suppressed by wind, dessication, low average temperature and large fluctuations of temperature from day to night. Woody species are scarce and if cut down for fuel are unlikely to regenerate for decades or longer. Trash and garbage disposal poses another problem. Bacterial and chemical decomposition tends to be negligible, so that litter accumulates and disfigures the landscape; it must be buried, burned or transported elsewhere. Prevention of trail erosion and the trampling of fragile plants by large numbers of hikers and climbers are other common maintenance problems, as is the excessive picking or collecting of flowers and other plants. The construction and maintenance of access roads that must climb through the high rainfall zone, which is likely to be uninhabited and covered by rain-forest, entail exacting engineering criteria if erosion is to be avoided, especially as land adjacent to such roads may attract settlement although it is unsuited to and liable to deteriorate under cultivation. Particular attention was paid to these points in a recent survey of Mount Kilimanjaro, the object of which was to evaluate the region's potential for tourism and the need for tourist access and accommodation. The analysis of economic benefits presented in this study (U.S. Department of the Interior, National Parks Service, 1970) should be of considerable interest to those responsible for planning the tourist development of similar areas.

(e) Prior Human Habitation of Potential Park Areas

Most of the remaining forested regions of the humid tropics are occupied by people engaged in hunting and gathering or shifting cultivation, although their populations may be very sparse. The presence of these groups in potential park ecosystems presents a dilemma which has little parallel in temperate zone park management experience. Such forest dwellers may be of extraordinary scientific and human interest, and in the context of park management objectives present unique moral problems. These problems have been more tractable and susceptible to rationalization in the context of settlement and exploitation of natural resources of virgin areas, however questionable may have been some of the actions taken to civilize 'primitive' inhabitants.

If, in the forests of the Amazon river basin, for instance, one recognizes the implicitly legitimate and prior claims of the numerous Indian tribes, it is likely that only the most uninhabitable (probably the most humid) forest is not included in the territory of one or other tribe. Similar considerations apply to the extensive forests of New Guinea and Malaysia and other large tropical forest areas. A whole series of fascinating and important questions concerning the establishment and maintenance of parks in these areas is suggested by this situation:

Should primitive tribes be permitted to continue their occupation of land designated as national park?

Should unique cultural groups be purposefully isolated from modern civilization or should they be exposed to its impact?

Would establishment of park boundaries stabilize relationships between previously competitive groups, eliminate warfare and result in population increases and ultimately in major alteration of the ecosystem?

Are primitive hunters and gatherers living in a park to be considered as part of the dynamic processes of the ecosystem?

These questions are not susceptible of standard answers, but they need to be asked by all park planners and authorities charged with establishing forest and biological reserves in the areas concerned.

The philosophical implications of these questions are not in fact altogether esoteric and may extend to other situations. For instance, should new 'labour saving' agricultural tools, derived from modern technology, be made available to people if the employment of these tools will cause irreversible disruption to an ecosystem?

Are those cultural groups widely admired for their values, traditions, cultural expression (language, arts, architecture) and apparently harmonious relationship with their environment to be isolated or protected from the impact of modern technology and contact with men from very different cultures? How is the worth of such cultures to be determined? Can it be determined? By whom? If, for instance, they are tourist attractions, is their value to be judged only by the number of tourists (e.g. market value) and the reasons for which tourists visit exotic cultural groups? Are such cultures to be considered as living museum pieces and objects of scientific interest?

For anyone who has questioned the adequacy of the values and social organization that have developed in societies based upon the scientific revolution (e.g. industrialized and technological societies), prescientific or non-scientific cultures possess great value for suggesting alternative modes of human and man/nature relationships. The importance attached to the work of Margaret Mead and other cultural anthropologists attests to the significance of non-scientific cultures for modern man.

(f) Marine Parks

The drawing power of marine parks for the growing number of skin and scuba diving aficionados is very high. A good example is the Buck Island Reef National Monument off the north-east coast of St. Croix in the Virgin Islands. Park visitation here jumped from 600 in 1962 (its first year in operation) to 15,500 in 1965 (Randall, 1969).

The feasibility of maintaining the environmental integrity of a marine park must be included in management considerations. Special regulations needed for such parks include control of shell and coral collecting, spear fishing and other sporting and commercial (or subsistence) fishing activities. Dynamiting of reefs is a particularly destructive practice that must be curtailed, especially in potential park areas. Pollution, dredging and erosion, leading to turbidity of the marine ecosystem, must likewise be brought under control.

In the planning and management of marine parks, pertinent questions are—

Is the proposed area accessible?

Is it free of pollution and sedimentation (e.g. sufficiently removed from population and industrial/port activities)?

Is the underwater life and landscape reasonably varied and beautiful?

Is the area relatively safe for swimmers (e.g. no strong currents, exceptional tides or dangerous surf)?

As in the case of terrestrial parks, features of archeological or historical interest (such as shipwrecks) may add value to potential marine parks. Shipwrecks or sunken vessels may be of great historical interest and the object of investigation of the growing field of underwater archeology. In the Americas particularly, they could constitute significant park attractions, many being located in relatively shallow water accessible to skin divers. The preservation of West Indian shipwrecks is one of the aims of the Caribbean Conservation Association. The possibility of viewing the underwater remains of a shipwreck can be a valuable asset to a marine park, and the establishment of wreck-strewn shoals or reefs as parks would have the double value of a tourist attraction and underwater archeological reserve. For the latter purpose, regulations prohibiting retrieval of artifacts by tourists would have to be strictly enforced.

(4) ECOLOGICAL ASPECTS OF COASTAL ZONE TOURISM

Coastal zones with benign climates have long been of major interest to tourists and the increasing popularity of marine water sports, particularly those underwater activities made possible by reasonably priced diving equipment, has made coastal areas prime targets for tourism development. The aesthetic and recreational attractions of tropical palm-fringed beaches, surf and colourful reef life to the tourist from the temperate zones are demonstrated by the emphasis on these features in travel advertisements.

However, the development of coastal zones and shorelines for tourism raises a series of environmental considerations concerning, in particular, public health (which has perhaps received most attention in the past), water quality, preservation of marine life, and maintenance of the ecological functioning and aesthetic values of the region. The marine ecosystems of coastal waters, which include lagoons, estuaries and the relatively shallow waters of the littoral zone, are perhaps the most complex, diverse and biologically productive ecosystems of the biosphere. They are also among the least understood, one of the reasons being that so many different disciplines are involved—for example, marine biology, reef ecology, geomorphology, climatology, oceanography and sanitary engineering.

(a) Sewage and Dredging Problems

In all coastal waters, but particularly in warm waters where reefs support a great diversity of marine life, the effects of sewage pollution and dredging require special attention and more research. Sufficient sewage to have a noticeable effect on coastal ecology would probably constitute a health hazard at the worst and be at best an unaesthetic nuisance. It is obvious that where sewage may pollute recreational waters, quality standards and monitoring problems are necessary to protect the health of swimmers. There are apparently no international water quality standards for water-contact sports. With respect to subtropical marine waters, researchers in the US Virgin Islands (Brody *et al.*, 1970) have recommended that in addition to faecal coliform bacteria, pathogenic streptococci be monitored and that selective tests for bacteria viable in seawater be undertaken, in order to yield a real measure of potential toxicity to humans. Maximum allowable coliform bacteria counts of 70/100 ppml were judged too high in the bay waters of St. John Island.

Other types of pollution are equally serious. As noted earlier, use of persistent biocides (as in mosquito control) may have serious detrimental effects on marine ecosystems. Oil leakage, spills and dumping are particularly harmful. Their immediate effects are to be seen in the destruction of coastal and marine bird life and in the contamination of beaches, making them unsuited for recreational use. Contaminants of industrial origin which may be of very considerable significance include polychlorinated biphenyls (PCBs) and heavy metals (lead, mercury, cadmium and arsenic). The practice of dumping solid wastes and debris is also a serious problem in many coastal areas and must be discontinued if the aesthetic values and tourist attractions of these areas are to be maintained.

Turbidity resulting from dredging and erosion can have disastrous effects on lagoon or bay waters, especially if wave or tidal action is not pronounced. Dredging operations not only totally destroy the dredged areas but also stir up fine sediments that may remain in suspension for up to a year and can reduce the sunlight reaching a shallow sea-floor by as much as 90 per cent (van Eepoel and Grigg, 1970). This will affect organisms living on the bottom, which include photosynthesizing plants such as turtle grass (*Thalassia*); if it results in the death of these primary producers and the consequent disappearance of a source of food and habitat, an entire spectrum of dependent life may be eliminated from bay waters.

Another consequence, perhaps even more serious, is the cutting off, through turbidity, of oxygen formerly supplied by photosynthesis to bay waters. All marine life sensitive to dissolved oxygen levels may then be adversely affected. Corals and other living organisms can also be killed by the physical action of sediments settling out from turbid water or deposited directly by dredging operations. In short, all such operations, as well as onshore earth moving operations associated with construction, must be undertaken with extreme care, in order not to destroy the very qualities of bay sites which make them attractive and desirable to tourism.

Estuaries may offer potential for development through landfill, particularly where suitable dry land sites are not available for development. However, the destruction of estuaries may seriously affect fisheries. Some fishes and crustaceans, especially certain shrimps, require variations in salinity for their development, which can only be found in estuarine areas. If few estuaries exist on a coastline (which is frequently the case with volcanic islands), their destruction through landfill could seriously alter fishery productivity.

Mangrove swamps play a key role as feeding and spawning areas for a great number of commercial and sport fish as well as many other marine organisms. In addition they have an important role in regulating the movement of fresh and salt water in the intertidal zone and provide a habitat for a number of different kinds of wildlife. Their protection is essential for the maintenance of the productivity of coastal waters.

In the Florida Everglades mangrove belt, detritus from red mangrove leaves is the principal source of food for the aquatic species of estuaries (Odum, 1971). The fallen leaves are decomposed by micro-organisms, which then are consumed along with the converted leaf detritus by higher omnivorous organisms such as small shrimp and midge larvae. These, in turn, are food for larger species. The destruction of mangrove forests would remove a major source of food for many animals and directly limit the productivity of these waters. Detritus-eaters are the prey of over sixty species of juvenile fish in Florida which live in the mangrove-bordered estuary for varying periods of their lives. It has been estimated that for every acre of estuary filled or dredged in Florida, two others are ruined for fish production by siltation, pollution and other disruptions (Robas, 1970, p. 21).

(b) Marinas

The tremendous popularity of boating and the great increase in the number of water-craft which cruise in the Mediterranean, Caribbean and Pacific territorial waters of Mexico and Central America, have resulted in an expanded demand for marinas. Ideal sites are, of course, coves and bays offering good protection from high seas and storms, but as mentioned above, these are the very areas in which the marine environment is especially susceptible to the damaging effects of pollution and dredging. The best sites are naturally also likely to attract or be associated with hotel and related beach development; but major developments of this kind in bay ecosystems, where the rate of natural flushing action by currents and tides is low, will usually, in the absence of efficient installations, lead to serious water pollution. The final outcome could well be bay waters largely deprived of marine life but full of noxious odours.

Sites which are subject to sedimentation from wave action or stream erosion could greatly increase the public or private cost of marina operation. The maintenance of navigable waterways, of coastal navigational aids such as buoys, and of navigational services such as charts and marine radio services, is usually at public cost. The provision of these services is indispensable to recreational boating and off-shore cruising, and must be considered as a necessary item in any plan which encourages or finances these activities. Much the same applies to measures aimed at attracting cruise ships. In Haiti, for example, cruise lines were encouraged in 1970 to arrange calls at Cap Haïtien on the north coast, where the bay entrance and the bay itself are infested with treacherous reefs and the channel near the port is subject to continuous sedimentation from a nearby river. Major effort and investment were required to improve the buoy system and dredge the channel before deep draft cruise ships could be docked and the Haitian Government itself had to put the work out to contract.

Construction of marinas and associated facilities in bays and estuaries, therefore, needs to be planned with a view to preventing pollution and to possible effects on sedimentation and erosion. Sanitary engineering and ocean engineering expertise is essential to the development of sound plans.

(c) Conservation of Unique Sites

The demands of increasing numbers of tourists attracted by the natural environment of coastal areas should be a major consideration in development planning. These natural features include underwater life, coastal land forms (such as caves, blow holes and waterfalls) and undisturbed coastal forests, with their wealth of flowering species, bird and other animal life. The attraction of unique or unspoiled sites cannot be overestimated. While many tourists may not fully appreciate the natural history value of such sites, their beauty is a great asset to a region's resources, which needs to be carefully safeguarded as an integral part of tourist development.

The protection and management of such attractions is urgent where facilities for tourists (e.g:, accommodations, shops, etc.), and accompanying local settlement (e.g., new housing for employees and residents) endanger the natural site. Privately financed development often actively seeks hotel locations as near as possible to outstandingly beautiful and unique coastal land forms and beach areas.

For this reason, it is vital that an inventory and the preparation of a zoning and management plan to ensure the preservation and public use of valuable natural sites should precede development. This recommendation is particularly relevant to situations where there is a high present or prospective pressure on coastal lands for recreational use, or where outstanding attractions are found close to existing high-density tourist facilities.

An example of a unique coastal area now threatened by shoreline development is provided by the phosphorescent bays of Puerto Rico. The luminescence of the waters of these bays is generated by acquatic microorganisms, and is particularly well developed along the south coast of the island, where it is a night-time spectacle enjoyed by many visitors. The biological balance of one of the bays has already been disturbed by a canal project, and the intensity of luminescence has declined. Increasing amounts of raw sewage originating from coastal homesteads, houseboats and industries also threaten the ecology of these days. An analysis by the US National Park Service of the impact of development makes out a clear case for the conservation of these remarkable bays as parks (US Department of the Interior, 1968).

The preservation of coastal sites has numerous precedents. Among the best known are the 77,000-acre John Pennekamp Coral Reef State Park on the Florida coast, the recently inaugurated 10,000-hectare Santa Rosa National Park on the Pacific coast of Costa Rica, the Virgin Islands National Park on St. John Island and the 176 square mile Exuma National Land and Sea Park in the Bahamas, which is 95 per cent under water. The establishment of a series of underwater parks along the East African coast is also partly completed.

The pristine beauty of palm-fringed tropical beaches may also constitute the home environment of indigenous populations existing at a subsistence level and deriving their living from the sea and small garden plots. Development of such areas should take into account the possible conflicts which may arise between resource use for tourism and for the livelihood of local inhabitants. Attempts should be made to anticipate such conflicts, create mechanisms to harmonize the development of resources and, wherever possible, blend tourist development with the local culture and way of life.

(d) Scale of Development

Too many new buildings can stimulate excessively high densities of tourists and overwhelm environmental attractions. Hence, proper consideration should be given to the optimum scale of development in tourism planning.

The possibility of measuring the carrying capacity of beach areas has been suggested in one evaluation of tourism potential (Piperoglou, 1966). In this study of coastal Greece, beaches were considered to be the pivot of tourist development. Maximum or peak hour densities of people on beaches were based on observations in the Athens area. The data indicated that 50 per cent of the tourists in this area are likely to use the beaches on a 'peak' summer day. Hence the optimum overall capacity was established as twice that of the optimum daily utilization capacity of the beaches. The following basis of calculations was suggested:

Type of beach	Type of accommodation	Persons/m^2 of beach
Small bays	High cost	0·05
Large bays	Medium cost	0·10
Long beaches	Low cost	0·15

These densities apply to Greece, and are quite obviously conservative compared to densities on Copacabana beaches on a peak Sunday. Nevertheless, conservative estimates should be the point of departure for assessing the capacity of tourist hotels to be built at or near first-class beaches, especially if water pollution and sedimentation control may be difficult or prohibitively expensive. The control of effluents from tourist hotels may be more economical for a single large hotel than for several smaller ones. On the other hand, a large high-rise hotel may overwhelm the scale of the physical setting and detract from its beauty. There are other advantages of small-scale tourist development. Small hotels in Barbados, for example, employ more persons per tourist than do large hotels (Lundgen, 1968). This suggests that planning for relatively low densities of tourists in coastal areas dominated by beach attractions could (if adequate pollution controls are included in the small hotel development) not only serve the purposes of environmental conservation and aesthetics, but also produce greater employment benefits, thereby advancing development goals. In addition, a steady increment in hotel capacity would be relatively easy to plan through successive construction of small inns and hotels. This policy would avoid the administrative and employment problems created by the sudden appearance of very large hotels and would favour a rational and steady build-up of the reputation of a region and its facilities.

(e) Additional Safeguards in Coastal Development

Although, in previous sections, emphasis has been placed on the identification of specially valuable sites and their protection—always the first priority—, the control of pollution and the need for special attention to estuarine and mangrove ecosystems, three additional methods of helping to ensure the ecologically sound development of coastal areas are worth noting:

1. Establishment of bulkhead lines at mean high tide level and prohibition of in-filling of tidal wetlands below these bulkhead lines can be useful in protecting valuable aquatic resources. Moreover, dredging should only be permitted below the bulkhead lines after surveys have been undertaken to identify productive areas for marine life. Thus dredging would necessarily be excluded from these areas and adequate measures taken to prevent siltation and increased turbidity from the dispersal of suspended solid material in the tidal waters of the productive zones.

2. To avoid salinization of freshwater aquifers, dredging of sea-level canals in estuarine regions should generally be restricted to the area normally reached by the tidal flow of saline water. Establishment of a legally determined 'salinity line' beyond which such dredging is prohibited is one means of accomplishing this.

3. If such adverse consequences of development projects as beach erosion and the siltation of navigable channels are to be prevented, it is

essential that in planning the construction of docks, piers, marinas, canals and other coastal facilities, the most careful attention is paid to coastal currents and the way in which silt loads carried down to the sea by rivers, streams and runoff are consequently distributed.

(5) ECOLOGICAL ASPECTS OF TOURISM IN ISLAND ENVIRONMENTS

The use of islands for tourism is a relatively recent human activity among many others which have so greatly altered these fragile ecosystems. To consider islands in the context of tourism brings into sharp focus their ecological vulnerability to human alterations. The ecological integrity of an island landscape is liable to be a measure of its attractiveness. The more 'unspoiled' it is, the greater is its potential for tourism. The epitome of western man's dream of 'getting away from it all' is a carefree existence on an untouched tropical isle, with abundant sunshine, fruits, flowers and guileless young maids. The persistence of this simple recipe for utopia in the imagery of travel advertisements is testimony to the rather specific nature-oriented attraction represented by islands. However, in reality, the development which takes place to accommodate the tourist can destroy or greatly alter the natural beauty and ecological balance of the island ecosystem, if in fact men's earlier activities have not already done so. The previous discussion of tourism development in coastal zones is applicable to islands as well as continents and the present section will therefore emphasize the effects on terrestial island ecosystems.

Because of their isolation and often their relative geological youth, island flora and fauna do not display the diversity or resistance to biological invasion encountered on continents. However, extraordinary evolutionary radiation in form and habit is commonly found in the endemic flora and fauna of many oceanic islands. The ancestral stocks may have been very limited, but have evolved into numerous species suited to different niches in the island ecosystems. It is this remarkable process as observed in the Galapagos Island which so greatly influenced Charles Darwin's thinking on the evolution of species. The endemic flora and fauna of the Hawaiian Islands similarly exemplify extensive speciation among relatively few genera, adapted to narrow geographical and ecological limits. It results in great vulnerability to competition from introduced exotics.

The impact of man and his domestic animals on island ecosystems has been tremendous. Eighty per cent of the native Hawaiian land birds have become extinct, especially ground nesting species which succumb easily to introduced rats, mongooses, dogs, cats and swine (Zimmerman, 1963). Forty per cent of the world's vertebrate extinctions have occurred on the islands of the Caribbean (Carlozzi, 1971). Fire, land clearing, and the browsing habit of goats have resulted not only in the disappearance of the native vegetation on many islands, but also in accelerated soil erosion. Finally, the problem of population growth and numbers of people in excess of carrying capacity is dramatic, notably on many smaller islands with relatively poor resources.

The introduction of exotic plants and animals to island environments poses a special threat to island life forms. It should on no account be undertaken on any island on which distinctive endemic flora and fauna still exist, unless there has been prior study of the potential of the introduced species to spread into undisturbed natural areas. When introductions are deemed essential, every effort should be taken to prevent the spread of the introduced species into natural areas. Where exotic animals, or plants have become established and harmful to native species, their eradication or control should have a high priority in development planning.

The continued existence of feral goats on Guadalupe and in the Galapagos, for example, threatens with extinction many native species that still survive and has already brought extinction to several others. The procedure adopted for eliminating feral animals from the Channel Islands of California and from the Hawaiian national parks may be a useful guide in dealing with this problem.

Tourism is capable of a major role in the economic development of islands. Apart from the occasional ones such as gambling and tax free shopping, the tourist attractions of an island environment are its climate, the beauty of its landscape and its plant and animal life. The development of tourism normally gravitates to these natural features, perhaps supplemented by the attraction of historic sites. The unique cultures of many island populations may be an added attraction, the more 'exotic', the more powerful. From the standpoint of the tourists alone, all these attractions should certainly be conserved, but unfortunately most aspects of tourism development are antithetical to cultural conservation and the protection of unique environments, since they imply modernization with resulting cultural changes, urbanization and resource exploitation. The problem is to reconcile the commonly conflicting aims of development and the conservation of natural attractions.

Carlozzi (1968) develops a strong rationale for the conservation of nature and of historical sites in the Caribbean as an integral and necessary part of tourism, which he considers to be a major opportunity for achieving a measure of economic self-sufficiency in the Lesser Antilles. His proposals to accomplish an integrated tourism development and conservation programme include:

inventory of natural and historical assets;

land acquisition of beaches to preserve public access;
restoration and stabilization of historic sites;
construction of access roads and facilities for potential park sites;
formation of foundations and trusts, in the absence of institutional capabilities and in view of the specific nature and small scale of site preservation;
technical assistance in park and recreation planning, restoration, museum design, interpretive programmes and scientific research;
citizen education in environmental and historical site conservation; and
formation of a Caribbean international park system, aimed at establishing common standards and coordinating park development.

Since most oceanic islands have an unusual combination of scenery and biota, they offer opportunities for appropriately planned recreation and tourism less often found in mainland areas. They have a special appeal to visitors, who would be less likely to spend equivalent time and energy on visiting mainland sites. However, much of this appeal can be lost by poorly planned or unplanned development resulting in the spreading of more or less uniform hotel-resort facilities over areas that would otherwise be different. In many cases the encouragement of scientific and educational tourism based on the special natural and cultural conditions to be found on certain islands may prove more economically advantageous to these islands than equivalent expenditure on attracting mass tourism.

Certain islands may have such unique biota and be so vulnerable to human occupation that they merit total preservation in their natural state. The Galapagos Islands obviously fall in this category. These islands 'literally bulge with biological curiosities, including arborescent cacti and composites, four-eyed fishes, ocean-venturing iguanas, gargantuan land turtles, flightless penguins and cormorants, blood-eating and tool-using finches, sedentary seals and sea-lions' (Bowman, 1966). The 100 sq. km national park designated by Ecuador on a portion of Santa Cruz Island under the cooperative supervision of the Charles Darwin Research Station is a welcome step, but much of the unique endemic fauna of the islands remains threatened by poachers, domestic animals and most recently by large numbers of tourists.

Milton (in Porter *et al.*, 1970) describes the relatively recent concern over the destruction of the Galapagos ecosystem and its unique species, and traces the international efforts since 1957 which led to the establishment of the Charles Darwin Foundation and the location of its research station on Santa Cruz. He notes the possibly paradoxical role that tourism can play in this uniquely beautiful—and therefore fragile—environment. By providing alternative sources of income to local inhabitants of the Galapagos, tourism can indirectly diminish the predations and intrusions of man and his animals on the ecosystem and its species. Excessive numbers of visitors and uncontrolled collecting could, however, have an equally destructive impact on the archipelago, and 'smother its beauty' by too much attention. A middle course is required in which the tourist carrying capacities and other problems of the various islands would be defined through appropriate research and conservation practices applied by an informed and competent park management and with the support, based on education, of the local population.

The atoll of Aldabra and, also in the Indian Ocean, Cousin Island in the Seychelles, are other examples of islands whose remaining endemic flora and fauna merit preservation; both now have biological reserve status. Relatively undisturbed islands are liable to be so rare that first priority should be given to their value for science rather than for tourism or other uses. In some cases, both scientific values and park-based tourism are compatible. On the island of Dominica, which has the only large extent of tropical rainforest left in the Lesser Antilles, the Conservation Foundation has recommended to the Government the preservation of a portion of this forest as a national park: it would have the multiple value of watershed protection, of appeal to tourists who seek the qualitative experience of visiting a rainforest, and of a scientific reserve for parrots and other species in danger of extinction (Conservation Foundation, 1970).

The conservation of insular flora and fauna has been championed mainly by dedicated scientists, but the great interest of islands and their demonstrated ecological vulnerability combine to emphasize the value to mankind of protecting the few remaining, relatively untouched island environments. Although undisturbed islands or parts of islands may appear to have a great potential for tourism, it would be unreasonable not to give prior consideration to their scientific value as possibly unique sites. In the long run, the capability of such areas to attract and cater for visitors will surely be enhanced and suitable parts could then be protected and managed as national parks or equivalent areas.

Meanwhile, oceanic islands that remain uninhabited by man and have not yet been devastated by human activities should be seriously considered for inclusion in the network of 'islands for science' which it is planned to establish under an international convention. Such consideration should be extended to include relatively undisturbed portions of larger, inhabited islands with a view to the establishment of similar scientific reserves, partial reserves (for the protection of breeding colonies of particular species) or, where appropriate, national parks. Such protection, through parks and reserves, of island biota of exceptional interest, including intertidal and subtidal zones, would constitute a most significant step towards the rational use and conservation of island environments.

(6) REFERENCES

Anon (1960). The distribution of tourist expenditure in Jamaica. *Revue de Tourisme*, Jan.-Mar. 1, pp. 22–7.

Bowman, Robert I. (1966). *The Galapagos; Proceedings of the Symposia of the Galapagos International Scientific Project.* University of California Press.

Brody, Robert W. *et al.* (1970). *Study of the Waters, Sediments and Biota of Chocolate Hole, St. John, with Comparison to Cruz Bay, St. John.* Caribbean Research Institute, St. Thomas, Virgin Islands.

Carlozzi, Carl A. (1972). An ecological overview of Caribbean development Programmes. In *The Careless Technology: Ecology and International Development.* John P. Milton and M. T. Farvar, eds. Doubleday and Co., Natural History Press, New York.

Carlozzi, Carl A., and Carlozzi, Alice A. (1968). *Conservation and Caribbean Regional Progress.* The Antioch Press. 151 pp.

COAP (1970). The Comprehensive Ocean Area Plan is under preparation. The matrices are taken from: Sorensen, Jens C., October 1970. *Coastal Planning: The Impact of Use on Environment.* Master's Thesis in Landscape Architecture, University of California in Berkeley, Graduate Division. The matrices were employed as working documents in the elaboration of the Comprehensive Ocean Area Plan, undertaken under the Department of Navigation and Ocean Development, State of California.

Conservation Foundation (1970). *Dominica; A Chance for a Choice.* Washington, D.C.

Darling, F. F. and Eichhorn, N. D. (1969). *Man and Nature in the National Parks: Reflections on Policy.* The Conservation Foundation, Washington, D.C. 2nd edition. 86 pp.

Gasparovic, F. *et al.* (1971). *A Study of Environmental Conditions and Problems in a Countryside Region attracting Mass Tourism—The Yugoslav Adriatic Coast.* (Paper presented at the May 1971 Conference on Problems relating to the Environment, Prague). 19 pp. Mimeographed.

Hart, William J. (1966). *A Systems Approach to Park Planning.* International Union for Conservation of Nature and Natural Resources, Morges, Switzerland. p. 107.

Houston, Douglas B. (1971). Ecosystems of national parks. *Science,* May, Vol. 172, No. 3984: pp. 648–51.

IUCN International Commission on National Parks (1971). *United Nations List of National Parks and Equivalent Reserves.* J.–P. Harroy, editor. Hayez, Brussels. 2nd edition. 601 pp.

Lewis, Phillip H. (1966). *The Outdoor Recreation Plan.* Wisconsin Department of Resource Development, Madison, Wisconsin. (Wisconsin Development Series).

Lewis, Phillip H. (1964). Quality corridors for Wisconsin. *Landscape Architecture Quarterly*, January.

Lungden, J. (1968). Barbados tourist study: a study in economic and geographic adaptation. *Revue de Tourisme*, Vol. 23, No. 4.

McHarg, Ian L. (1967). An ecological method for landscape architecture. *Landscape Architecture*, Vol. 57, No. 2.

Odum, William E. (1971). *Pathways of Energy Flow in a South Florida Estuary* (see Grant Technical Bulletin No. 7). University of Miami, Miami, Florida, 162 pp.

OECD (Organization for Economic Cooperation and Development) (1970). *Tourism in OECD Member Countries, 1970.* OECD Publications, Paris.

Piperoglou, John (1966). Identification and definition of regions in Greek tourist planning. *Papers of the Regional Science Association*, Vol. 28, pp. 169–76.

Porter, E., Milton, John P. and Brower, K. (1970). *Galapagos: The Flow of Wilderness, Vol. 2 Prospect.* Sierra Club and Ballantine Books, San Francisco and New York.

Randall, John E. (1969). Conservation in the sea: a survey of marine parks, *Oryx*, Vol. 10, No. 4.

Ray, Carleton (1968). *Marine Parks for Tanzania.* Conservation Foundation, Washington, D.C.

Ring, Ronald E., ed. (1970). *Oceanography in Florida, 1970.* The Florida Council of 100, Tampa, Florida.

Robas, Ann K. (1970). *South Florida's Mangrove-bordered Estuaries: Their Role in Sport and Commercial Fish Production.* (see Grant Information Bulletin, No. 4). University of Miami, Miami, Florida.

Russell, E. Walter (1970). *Management Policy in the National Parks.* Tanzania National Parks, Arusha, Tanzania. 24 pp.

Stone, Benjamin C. (1969). National parks as a national resource. In *Natural Resources in Malaysia and Singapore,* B. C. Stone, ed. Proceedings of the Second Symposium on Scientific and Technological Research in Malaysia, 1967. Kuala Lumpur. 265 pp.

Talbot, Lee M. (1966). *Wild Animals as a Source of Food.* US Bureau of Sport Fisheries and Wildlife, Washington, D.C. (Special Scientific Report No. 98).

US Department of the Interior, National Park Service (1968). *The Bioluminescent Bays of Puerto Rico: A Plan for Their Preservation and Use.*

US Department of the Interior, National Park Service (1970). *Kilimanjaro: a Survey for the Proposed Mount Kilimanjaro National Park, Tanzania.*

Van Eepoel, Robert F., and Grigg, David I. (1970). *Effects of Dredging at Great Cruz Bay, St. John.* Caribbean Research Institute, St. Thomas, Virgin Islands.

Withington, William A. (1961). Upland resorts and tourism in Indonesia: some recent trends. *Geographical Review,* **51**, 418–23.

Zimmerman, Elwood C. (1963). Nature of the land biota. In *Man's Place in the Island Ecosystem: A Symposium,* R. R. Fosberg, ed. Bishop Museum Press, Honolulu.

CHAPTER 6

Agricultural Development Projects

(1) INTRODUCTION

The purpose of this chapter is to review ecological principles and related concepts for the planning and management of projects designed, in general, to increase productivity in agriculture and, in particular, to introduce and disseminate technological innovations, such as improved crop and soil husbandry methods, new crop varieties and new inputs (fertilizers, pesticides, disinfected seed).

Increased production may confer benefits on the environment, through improved soil and water management techniques—both essential to sustained productivity—or adjustments in land use which retire from cultivation areas such as steep slopes unsuited to intensive cropping. But techniques employed to increase production may also generate unwanted damage, permanent or temporary, to the environment. Thus, ecological complications will often accompany large increases in crop yields. The positive effects of improved crop and soil husbandry are well known, but the possible negative effects deserve more attention in project planning as well as management.

(a) Agricultural Project Planning

Agricultural development projects may have one or a combination of the following objectives:

to increase yields on presently cultivated lands;
to bring into production new crops and new varieties;
to bring into production new lands for farming through clearing or irrigation; and
to adjust land use so as to conserve renewable resources more effectively.

Within this framework, physical resources are assessed in terms of the best yields known to be possible using modern production technologies for ecologically suitable crops. Ecological suitability is usually defined only in terms of agroclimatological and soil variables known to influence a crop. In turn, the crops are evaluated in terms of production and market economics to determine the expected returns on the investment and serve as a basis for loans to producers.

A well thought out project does not ignore ecological factors, although the emphasis is on plant–soil–climate relationships which limit or favour attainment of the best yields. If soil surveys precede project implementation, soil erosion danger is normally included as one of several elements in the evaluation of land capability and in recommending soil management practices.

The agronomic component of agricultural project planning seeks answers to the following questions:

What crops can be produced in the project area, or conversely, where can the desired crop be produced?
What are the productive capabilities of the soils in the project area, or in the climatic region suitable for a desired crop?

A comparison of present land use in high potential zones would then make possible the quantification of the project acreage, and indicate where land use changes will be sought to accommodate a new crop. Knowledge of the land tenure of the project area, including number and size of farms, would also be determined in order to plan agricultural credit and supporting technical assistance.

The technical feasibility analysis is complemented by an economic analysis of production costs, including labour, involved in soil and crop management, seed, fertilizer and pesticides. However, the last two may not be susceptible to accurate costing. Only detailed knowledge of soil characteristics in the root zone would permit estimates of fertilizer needs and, ultimately, only fertilizer response in actual cultivation will indicate the requirements. The need for pesticides and other chemicals to combat crop pests and diseases can be anticipated. but not accurately costed. However, generally speaking, pest and disease control costs tend to increase over time for a number of reasons related to insect ecology and population dynamics, and to impermanency of resistance to disease, which are discussed in Section 2(a) below.

The economic feasibility analysis for an agricultural project is based on answers to the following questions:

What are the production costs on the farm?
What is the foreseeable demand for a crop and its market price?
What are the transportation costs from farm to market?
What are the associated processing and storage costs, if any?

Information obtained from these questions will enable an estimate to be made of the viability of the farm operation and the investment as a whole.

The productivity goal of the project may be set at less than maximum possible yields, if factors external to the farm do not appear favourable to attaining them. Pertinent factors may include road networks, market location and functioning, extension services, educational level and experience of farm operators, land tenure arrangements, income elasticity, and social and cultural influences related to individual incentive. The allocation of capital resources to remove obstacles presented by these factors would not usually be made through a production-oriented investment, such as loans to farm operators for seed and fertilizer, but rather through institutional programmes, such as agricultural research and advisory services, or government intervention in marketing, land reform, etc. It is noteworthy that almost all the ecological problems described in this chapter require institutional actions and programmes of regional scope for their solution.

A consideration that applies to large scale projects is the impact of more efficient production techniques on the rural labour supply. This can lead to displaced labour and inability of small farms to compete in the marketplace. Although new technologies may also generate a demand for additional jobs in agriculture, displacement of farms or jobs is clearly a matter which can have repercussions on the movement of people from country to town as well as on the use of land resources, and is therefore of major concern. While planners are increasingly taking into account these socio-economic impacts, the ecological impacts need more attention. The spread of subsistence farming and overpopulation in regions whose resources are ill-suited to intensive production may result indirectly from more efficient production in areas with better physical conditions.

The language and concepts of economics and technology tend to obscure environmental changes associated with the diffusion of high-yield production methods. In fact what is happening is that man is reorganizing the production of plants and animals in an ecosystem. A limited number of species are produced on a large scale and there is an interruption in the successional evolution of vegetation towards a climax forest in which most of the energy exchange is for the purposes of maintenance and the production of cellulose. In an agroecosystem, on the other hand, much or most of the photosynthetic process is channelled into the production of carbohydrates and protein.

(b) The Relationship of Ecology and Economics in Project Planning

The environmental changes associated with an agricultural project are no less complex than the economic results, and are perhaps less predictable. The relationship between economics and ecology concerns, in effect, the interactions of an economic system and a biological life system, which in this context is the agroecosystem.

Figure 18 portrays the systems relationships between ecology and economics in the planning for and production of a single crop. The ecological side of the interacting systems is the larger biological life system whose basic components are nutrients, energy, water, species present and human populations. This constitutes both the natural resources base for agricultural production and also the larger habitat (or ecosystem) in which the production takes place, and which is affected by production changes. The disciplinary inputs pertaining to agriculture on the ecological side are the numerous branches of the earth and life sciences. The diagram shows only four of the major disciplines related to crop production: climatology, soils science, plant science and entomology. The first two are employed in surveys which establish production potentials. Research in soils, plant science and entomology establishes the agronomic requirements and potentials for crop production. Together, they establish the technical feasibility for the production of a given crop or the introduction of new production factors.

The economic side of the interacting systems is concerned with the economic feasibility and financial viability of production at the farm level. Within the larger economic system is the agricultural sector, the main

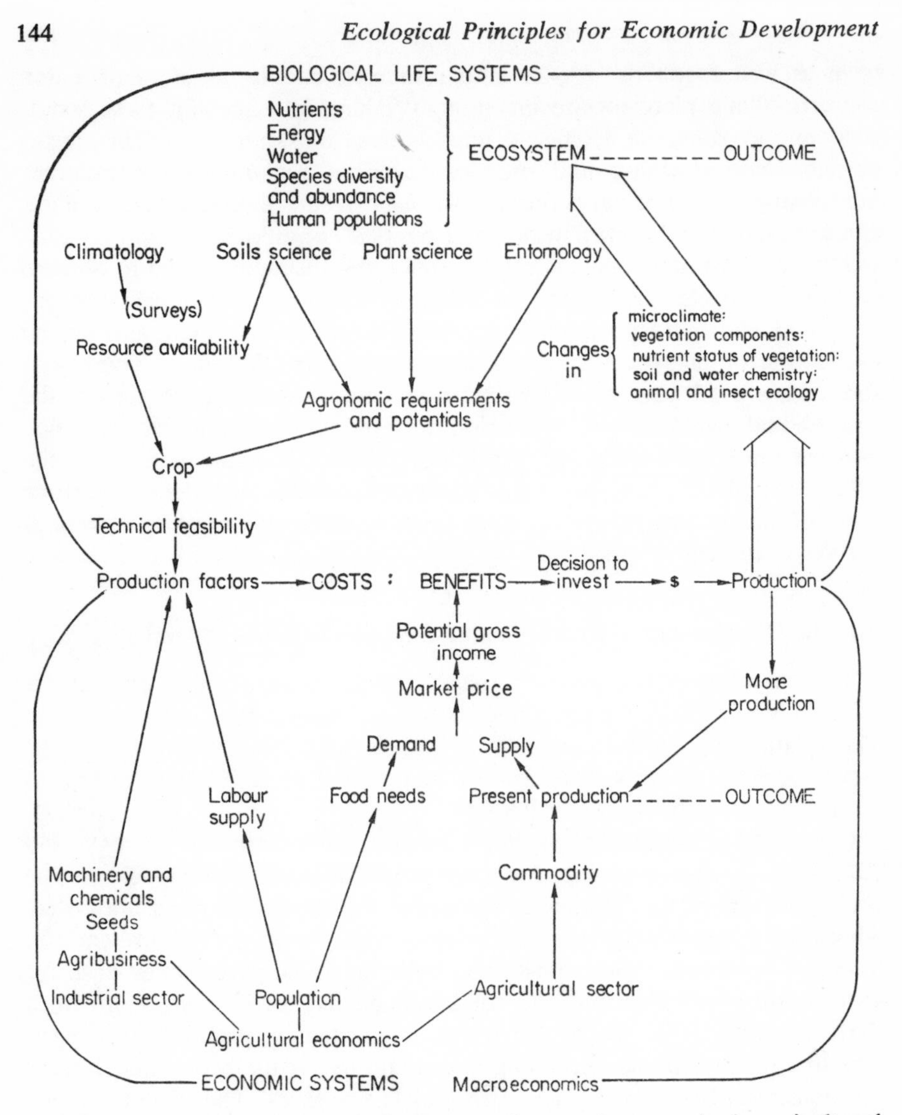

Figure 18. A systems relationship between ecology and economics in agricultural production.

concern of agricultural economics. The agricultural economist who is in charge of project preparation and analysis takes into account production inputs originating from the industrial sector (e.g. agricultural chemicals) and also labour supply, both of which are among the factors and costs of production. Analysis of demographic characteristics provide, in addition to estimates of labour supply, estimates of the demand for commodities which in this illustration is a food crop whose internal demand can be projected from growth rates, income elasticity and other data. The present production of the commodity represents the supply level. The present and future supply/demand relationships establish present and future prices, which in turn allow potential gross income from production, or benefits at the farm level, to be estimated. If the cost/benefit analysis and rate of returns are favourable, the decision to invest is made and the project gets underway. Production of the commodity then alters the supply picture, and sets off a chain of events which are fairly well understood by agricultural economists and are built into the analysis to the extent possible.

Changes in the agroecosystem accompanying the increase in production or use of new technology are numerous and may or may not be favourable. The objective, of course, is to attain stable and sustained production levels of the commodity. Whether or not this comes about depends in large part upon the new ecological relationships which accompany changes in microclimate, in vegetation composition (a new crop replaces others or is grown in previously forested land), in the nutrient status of the vegetation (food crops have high concentrations of nutrients), in soil chemistry and structure, on water chemistry and in animal and insect populations, and other changes arising from new or increased crops and the use of agricultural chemicals.

(c) Agriculture in the Tropical Environment

Developing regions and countries are by no means limited to the tropics and subtropics, but certainly some of the more perplexing problems of agricultural development are to be found there and, especially, in the humid tropics. Some of the ecological problems have been discussed in Chapter 3, but a brief restatement of salient features influencing agricultural development in the tropics may be useful as background for the detailed review of impacts, which follows in the next section.

(i) In the tropics, most lands susceptible to agricultural development are already under some form of use, including subsistence cultivation or pastoralism. Few projects will be aimed at virgin land, but many may affect mature forests or other ecosystems which would be more valuable for non-agricultural uses, such as maintenance of gene pools, scientific biome studies, forestry, wildlife habitat, national parks and watershed protection (Figure 19). Several of these uses are not, of course, mutually exclusive.

(ii) Biological processes in the conditions of continuous high temperatures and high humidity are accelerated. Tropical forest has been used as a measure of photosynthetic rates and, compared to temperate hardwood forests, draws three or four times the quantity of chemical elements into the biological cycle every year. Moist tropical forest also shows a greater

Figure 19. Upland sites, often with rich volcanic soils, attract agriculture but vulnerability to erosion can make agricultural development risky. Too often high values for land stability, science or tourism are lost when such areas are subjected to shifting agriculture. Parc National des Virunga, Zaïre. (Photograph by Jacques Verschuren: courtesy I.N.C.N.).

return of chemical elements in litter than any other type of ecosystem. Biological and chemical breakdown of organic matter and minerals is very fast, as evidenced by the rapidly declining yields of cleared areas. The tropical climate also speeds the life cycle of insects and other higher animals.

(iii) Tropical ecosystems display greater biological diversity than any others.

(iv) Given sufficient moisture, the growing season in the tropics is uninterrupted. This makes possible the double and triple annual harvesting of grain crops. For subsistence and commercial farmers, it means that agricultural activities go on throughout the year and that harvesting and planting operations may occur simultaneously.

(v) Regional differences in ecological conditions affecting agriculture to a large extent depend on humidity, that is the combined effect of rainfall and temperature, a general measure of which can be obtained by comparison of potential evapotranspiration (P.E.T.) to actual evapotranspiration. A useful gauge is the ratio of effectively dry months (P.E.T. greater than rainfall) to effectively wet months (rainfall greater than P.E.T.). Table 2

Table 2. General environmental and agricultural conditions according to length of dry season in tropical latitudes

	Six effectively dry months	Two effectively dry months or less
Solar energy	Higher	Lower
Ecosystem stability	Unstable: rainfall variable; biological life cycles fluctuate seasonally	Stable: less fluctuation of biological life; processes continuous
Soil characteristics affecting farming	Toxic salts; drought; high erosion danger; high pH.	Inadequate drainage; laterization danger; leaching of nutrients, loss of organic matter; low pH.
Pests and diseases	Seasonal explosions of insect and bird populations, migratory populations	Fungus, virus, bacterial diseases; beetles, ants, small mammals important pests; Many interconnections in insect, bird and mammal life
Weeds	Seasonal—not a major problem	Major problem
Cultivation (labour) schedule	Seasonal	Less seasonal
Important subsistence food crops (not grown under irrigation)	High protein grains: millet, sorghum, maize[a]	High carbohydrate tubers and grains: cassava, sweet potato, taro, rice
Important properties of food crops	Drought resistance	Low fertility tolerance (except rice); store well (tubers)
Secondary food sources	Plants: relatively few; but domesticated browsing and grazing animals may be important	Plants: numerous; wild game may be a source of animal protein
Risk of crop loss	High	Low
Likelihood of human overpopulation	Greater	Lesser
Major food supply problems in condition of overpopulation	Famine	Malnutrition (protein deficiency)

[a] Maize is one of the few grains that can be grown under a wide range of temperature, rainfall and humidity conditions, but is not a dominant crop in the very humid tropics.

summarizes the differences in ecological and other characteristics for two humidity extremes in the tropics.

The major technical problems of agricultural production in the tropics are control of weeds, diseases and pests; maintenance of soil fertility; and the provision of optimum amounts of water in the root zone, whether through drainage—where rain is excessive—or irrigation. Protection of the harvested crop from pest damage is a related problem. These are not at first glance much different from the problems of agriculture in temperate zones. However, they are in fact compounded by the rapidity of biological and chemical reactions and life processes, which are also the basis of high agricultural and plant productivity, and by the natural biological diversity of the tropical and subtropical zones. The informed tropical farmer will try to take advantage of biological diversity and the associated natural regulatory mechanisms on weed, insect and other pests, while at the same time managing crops and soils to obtain useful products. Thus he himself becomes one of the regulatory agents contributing to an artificially maintained equilibrium in the agricultural landscape. Many of the ecological problems in tropical farming revolve round the maintenance of this equilibrium.

(d) Soil Productivity and Management Considerations

The productivity problems of the soil resource involve the full range of technical, economic and social forces operative in agricultural development projects and their study essentially requires an ecological approach. Soil characteristics are an expression of climate, vegetation, relief, parent material, time and man's activities. The productive potential of soils is determined by the influence of these factors and the state of knowledge in soil and crop husbandry applicable to various soils. Attainment of potentials is further influenced by the financial and economic viability of different levels of farm management and farm size under different crops and intensities of land use. Similarly, different interpretations of property and different types of land tenure influence the economics of production and even the choice of crops, and so constitute yet another factor in the feasibility of attaining productive potentials. The relative significance of rural population densities and growth rates is determined to a large extent by the potential of the soils in a region, namely their capacity to support farming communities at varying levels of technology.

An appraisal of the productive potential of soils is an essential ingredient in determining the types of crop and technical input to be encouraged in a development project. The accurate determination of soil potentials is especially important in regions with dense populations of subsistence farmers, whose capacity for increasing production must depend mainly on innovations that produce higher yields from ground that at present barely provides for the needs of a family.

The extent and geographical limits of agricultural projects designed for specific areas will frequently correspond to the boundaries of soils which respond well to planned technological input. Conversely, soil problems which are expensive or technically difficult to counter can constitute basic limiting factors in the scope, objectives and production goals of agricultural development projects. Thus there is a direct technical relationship between agricultural projects and soil productivity potentials (in the technical analysis, defined initially in terms of maximum yield under optimum soil and crop management), although it will be modified by the range of demographic and socio-economic considerations mentioned above.

(2) THE ECOLOGICAL IMPACT OF AGRICULTURAL PRODUCTION TECHNOLOGY

The word 'impact' covers the rapidity, importance or magnitude of the changes associated with dynamic natural mechanisms put into play following manipulation of the environment. These mechanisms operate to bring the affected ecosystem into a new state of adjustment or relative equilibrium, which in the case of ecosystems dominated by agricultural land use, is an artificially maintained equilibrium. A major alteration in the vegetation cover induces shifts in organisms which depend, directly or indirectly, on plant life for food or habitat. The emergence of pests and diseases following the introduction of new crops or more productive varieties is a natural ecological consequence of their cultivation and one of the major preoccupations of modern agriculture.

At the same time that the 'economy of nature' is being reorganized for man's greater benefit, changes in human economic and social organization and functioning occur, whether as an objective of the development project or as a consequence of its implementation. These changes may affect the environment in both beneficial and harmful ways. Seldom do they have no effect whatsoever. Credit and price incentives, which encourage the cultivation of a crop better suited to soil and climate conditions than a traditional crop, exemplify an economic stimulus which produces environmental as well as economic benefits for the producer. The deforestation of erodable or quickly degraded soils in order to grow a crop which is temporarily in high demand represents a deleterious environmental effect of an economic change. In general, most of the environmental damage and ecological complications produced indirectly by economic stimuli, occur when benefits to the producer are perceived to be very high.

To the extent that new production techniques may have adverse effects on the environment or create unstable conditions for crop production, socio-economic and environmental costs are generated. The rest of section (2) discusses in detail the ecological outcome of employing certain agricultural production technologies that are known to create complications or inflict damage on the environment. The discussion should help the planner to understand how negative effects and their costs are produced, and how to avoid them.

(a) The Ecological Implications of Pest and Disease Control

The toxicity to many forms of biological life of certain chemicals such as DDT and other organochlorines, and of organophosphorus and heavy metal (arsenic, mercury) compounds, has been established largely on the basis of research and observations in temperate zone countries. These chemicals are used as insecticides, fungicides and herbicides to suppress plant and animal pests, and unwanted weeds. Their use has become synonymous with modern, high productivity farming. It is looked on as essential to the protection of crops from pests and to the control of insects that carry human disease, such as malarial mosquitoes.

The ecological complications arising from attempts to control organisms which make inroads on crop and animal production have been summarized by Smith and van den Bosch (1967) as follows:

'The use of chemical pesticides without regard to the complexities of the agricultural environment has been in recent years a major cause of disruption. Often, the target pest species has become tolerant of the pesticides and no longer can be controlled economically with chemicals. The population of the target species may quickly recover from the pesticide action and for a variety of reasons may rise to new and higher levels. Other non-target insects may, following the pesticide treatment, increase in numbers to damaging levels. The pesticide chemical may remain in or on the crop, or in the soil, drift to other nearby crop areas, flow into streams and drainages, and thereby, create a hazard to man or his animals, or produce additional side effects. The pesticide chemical may create hazards to pollinators, wildlife and other beneficial forms.'

(b) Pest Resistance to Pesticides

Brown (1957) reported that 60 species of arthropods were known to be resistant to insecticides, more than half of them to DDT, more than a third to other organochlorines and a sixth to organophosphorous compounds. Almost half of the resistant species were of medical or agricultural importance.

Resistance is basically the ability of an insect to metabolize the chemicals concerned. It is produced from normal populations by selection, and is not acquired in the insect's lifetime. It develops faster when the selection pressure is higher (for instance, if 100 per cent control rather than 90 per cent control is sought). Also it appears that, except in the housefly (*Musca*), resistance is inherited even though selection pressure (e.g. application of pesticides) has discontinued.

The number of resistant insects has undoubtedly increased since 1957 and, as pesticide use increases, has become more generalized in the agricultural areas of the world.

A common reaction to pesticide resistance is to increase the amount and frequency of chemical applications, e.g. from every two weeks to every three days in the case of Canete Valley cotton (Boza-Barducci, 1972), and to switch to pesticide formulations for which the insects have not developed resistance; these are usually more toxic forms of persistent chlorinated hydrocarbons (e.g. dieldrin and aldrin), or of organophosphorus compounds (parathion and malathion). The latter rapidly lose their toxicity, but parathion is extremely hazardous to humans as well as other living things. As a result, crop production costs increase and the ecology of the insect populations, both beneficial and harmful species, is further disrupted. Selection pressures continue and it can be expected that resistant individuals will survive and multiply.

(c) Disruption of Pest–Predator–Parasite Ecology

The suppression of insect pests of a crop also affects their predators and parasites. Other insects which prey on the target species may be killed as a result of pesticide use. Birds and other vertebrates may in turn be poisoned by feeding on the affected species. As the target species declines in number so do its parasites. Thus, the natural controls are greatly diminished and when insecticide resistance develops in the pest population, its resurgence can be dramatic.

An additional complication is that an insecticide may create new pests. In Kenya, an outbreak of the giant coffee looper in coffee plantations was traced to the lethal effect insecticides had on its parasite (Bigger, 1969). Neither the giant looper nor its parasite were target species. Relatively moderate amounts of parathion had been applied to kill other insects and, previous to the outbreak, the giant looper was not considered to be a pest of any economic importance.

Figure 20 below illustrates some of the possible causes and effects associated with the application of an insecticide to a crop and the development of resistance to the chemical by its pests. The time factor is not indicated,

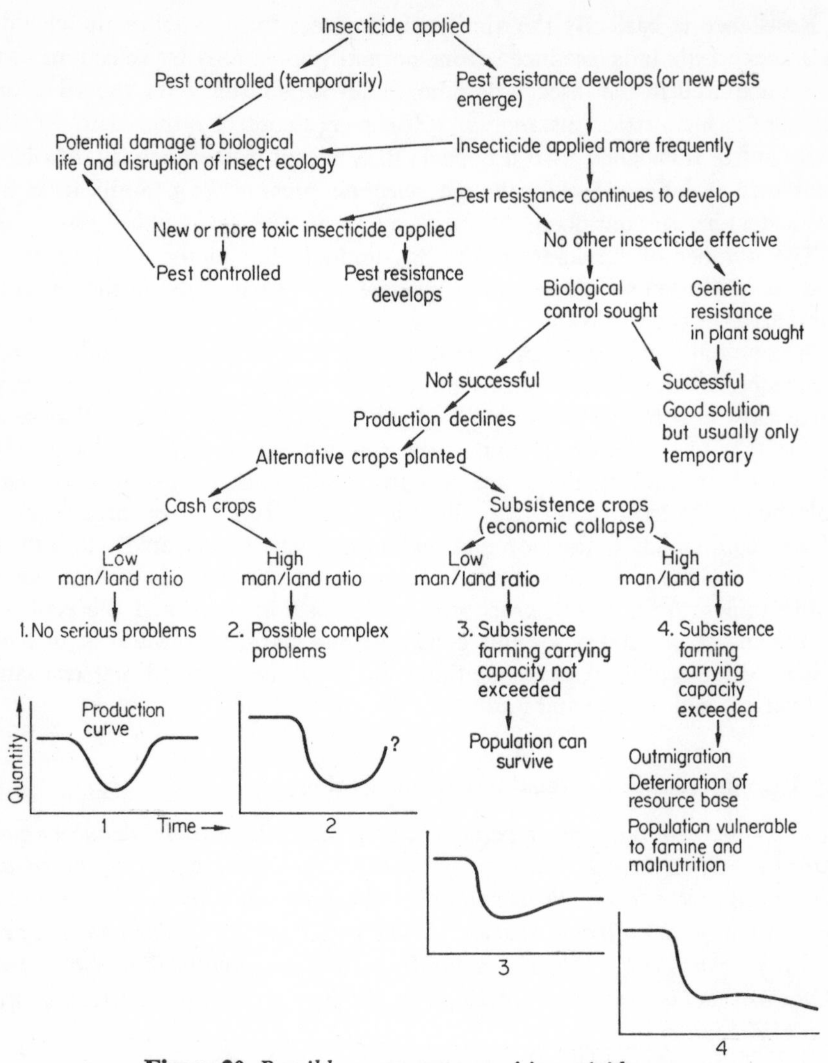

Figure 20. *Possible consequences of insecticide use.*

but based on the experience of agricultural scientists and entomologists with similar agroecosystems it should be possible to arrive at a reasonable estimate of the time lapse between initial application and the development of resistance, or population shifts, in pest species, or the collapse of disease resistance in a plant. Both events can be assumed to be probable and must therefore be taken into account in project planning. In the illustration, the importance of population densities and, implicitly, of farm size, becomes apparent when the worst possible outcome, e.g. reversion to subsistence farming, is projected. The potential for such an outcome argues strongly for maintenance of crop and vegetation diversity in the agroecosystem so that uncontrolled pest attacks do not bring about the total demise of a farming region. Past famines in northern Africa, resulting from locust attack on staple food grains, exemplify the vulnerability of subsistence farming populations, as does the more recent famine reported in northern Kenya, which was due to drought. Both regions are sub-humid and inherently unstable ecosystems.

Admittedly, the eventuality of a densely populated region being forced to revert from cash crop farming to subsistence farming (as shown under No. 4 in Figure 20) is an extreme hypothesis, but could possibly occur in certain rice farming regions. Collapse of disease resistance in high-yield rice varieties is perhaps more probable than, for instance, the development of resistance by the rice stem-borer to all pesticides. However, the catastrophic implications of switching from a major food grain such as rice to other, less productive food crops (both in terms of volume and total food value) underline the paramount importance of research in plant science and entomology, and of taking such implications into full account in agricultural project planning.

(d) Pesticide Contamination of Water

Persistent organochlorine pesticides, such as DDT, dieldrin and aldrin, are practically insoluble in water, but can be carried in suspension or absorbed in sediments. Only minute amounts of residues from this group of pesticides have been detected in water. Nevertheless these minute amounts, measured in parts per billion, are biologically magnified in food chains. In Lake Michigan, where DDT concentrations of less than 1·0 ppb in water have been found, some trout and salmon had 3 to 6 ppm, and fish-eating gulls up to nearly 100 ppm or roughly 100,000 times more than the water (Harrison, H. L. *et al.*, 1970).

Despite considerable research, knowledge of the long-term effects of these persistent chemicals in aquatic environments is still only superficial. The consequences of DDT and its breakdown products in aquatic ecosystems range from inhibition (at 10 ppb—Wurster, 1968) of photosynthesis of certain phytoplankton at the bottom of the aquatic food chain to declining reproduction of fish and oceanic birds. Aquatic invertebrates, especially insect larvae, spiders and crustaceans, are sensitive to very small amounts of DDT. Less than 1·0 ppm is lethal to shrimps. Mussels and snails, on the other hand, seem highly resistant. DDT is now found in the living tissues of mammals, birds and almost all fish from the Arctic to the Antarctic (NAS, 1971). It has been a major factor in the decline of some bird populations, notably in-shore species such as brown pelicans and cormorants, because of its now proven effect on reproductive metabolism leading to the laying of unhatchable, soft-shelled or shell-less eggs (Figure 21). Residues of polychlorinated biphenyls (PCB), derived from plasticizers used in paints, resins and many other industrial products, are also implicated, having been detected in several marine birds, especially again the brown pelican, but the precise effects are still undetermined.

Thus, the *magnitude* of the biological threat posed by the cosmopolitan occurrence and persistence of organochlorines in the marine environment is only partially measured. However, the *importance* of their environmental impact is indisputable. It is the subtle effect of non-toxic concentrations of persistent organochlorines or marine and fresh water ecology which is of greatest concern. Major disruption in species composition and food chains

Figure 21. Heavy use of persistent insecticides, particularly in agriculture, has had widespread effects. Thus severe breeding failures in birds of estuaries and inshore waters contaminated by runoff, have been traced to the accumulation of DDT or similar chemicals in their tissues. By contrast, the white pelican of S.E. Europe, W. Asia and Africa, shown here, has so far been more fortunate. (Photograph by Paul Géroudet: courtesy World Wildlife Fund).

is to be expected but, due in part to relative ignorance of the ecology of marine life, is still not susceptible to precise assessment. It could well have specially serious economic consequences for the highly productive estuarine waters, which are so important as spawning and fishing grounds.

With some exceptions, other classes of pesticide are not easily leached from soils, especially soils with high organic content, so that runoff contributes only minimal amounts to water bodies. Also their relatively low persistence diminishes the danger to the aquatic environment. However, major contamination can result from aerial spraying of pesticides deliberately applied to lakes and wetlands or allowed to drift over them. This can result from malaria control programmes and spraying of agricultural areas adjacent to lakes, streams and estuaries. Application of pesticides to rice-field irrigation water is another source of direct contamination. Lindane (gamma-BHC), which is employed in this manner to control the rice stem-borer, is lethal at concentrations needed for pest control to the fish *Tilapia mossambica.* This species is raised in rice fields throughout South-east Asia as a source of supplementary food and income. The herbicide 2, 4-D which is water soluble, can also produce fish kills. Herbicides are discussed further in subsection (g) below and the contamination of irrigation water in Chapter 7.

(e) Plant and Animal Uptake of Toxic Chemicals

The inadvertent contamination of food by toxic chemicals is a perennial danger associated with modern farming. The danger can be minimized by scheduling pesticide applications so that toxic residues will not be present at the time of harvest and by thoroughly washing the portions of the treated crop that are to be eaten. However, there is also a danger of pesticides being concentrated in the plant itself. The effectiveness of certain insecticides is based on the ability of plants to take up these chemicals. These are the so-called systemic insecticides, many of which are intensely poisonous to man (Negherbon, 1959) but widely used in small concentrations to control insects. They are applied as foliar sprays to the soil, or to the stem and trunks of plants, and are eventually translocated to the leaves. Before decomposition into less toxic substances they poison the insects that ingest the plant juices. The systemic pesticide dimefox is used to control the mealybug in coffee and cacao, and various other compounds to control the cotton aphid and the pests of such crops as sugar-beet, Brussels sprouts and cauliflower. Great care must be taken (and instructions are usually given on pesticide containers) to avoid the possibility of toxic residues affecting the edible leafy portions of plants treated with these systemic insecticides, and of discarded or unused portions, which may still contain chemicals that have not decomposed into non-toxic substances, poisoning domesticated animals and fowl that feed on them.

Some of the persistent insecticides have also been found capable of of penetrating and translocating in plants. In the United States, dieldrin has been detected in sugar-beet tops, corn, oats and alfalfa (Harris and Sans, 1969). Lindane is also absorbed (Reynolds, 1957), which is of particular significance in rice-growing regions where lindane residues from control of rice stem-borer may contaminate rotation crops, such as sweet potatoes. The avenues of contamination can be complex and subtle. In a Canadian instance, quoted by Frank (1970), the source of high concentrations of dieldrin in the milk butterfat from a dairy herd was traced to teak shavings and chips used for litter, which had been eaten by the cows when they were on a restricted diet. The teak chips, which contained dieldrin levels of 2 to 15 ppm, originated from logs grown in Colombia, South America, and shipped to Ontario.

In warm-blooded animals and man, in particular, organochlorines such as lindane and DDT are detoxified in the liver, but are also stored in body fat. Experiments with monkeys have shown that organochlorines stored in lipids are not toxicologically inert, but can be mobilized by stress and may then cause toxic responses in the brain. At doses lower than lethal or convulsive levels they can also affect the sensory processes of the central nervous system. For man, potentially hazardous levels of organochlorines in brain fat have been estimated 0·5 to 1·0 ppm for endrin, the most toxic, and 20 ppm for DDT (Revzin, 1970).

Residues in beef and dairy products are the main avenues for human ingestion of pesticides derived from animals. This is of concern in countries where the consumption of these products is relatively high. In Southeast Asia, where fish is a main source of animal protein, persistent pesticides may also enter the human body from this source. Fish raised in rice fields or inhabiting rivers, deltas and other waters contaminated by organochlorine insecticides, are liable to have ingested and stored these chemicals.

(f) Integrated Pest Control—an Ecologically and Economically Sound Alternative

In view of the undesirable and damaging effects on the environment of many pesticides, especially the persistent organochlorines, including DDT, dieldrin, aldrin and endrin, agricultural projects should be planned so as to minimize or obviate the need for these chemicals as production inputs. Planning should aim at minimizing the exposure of humans and other non-target organisms to the highly toxic, albeit not persistent, organophosphorus pesticides. If these aims are to be achieved, disease and pest problems must be anticipated and control measures used which employ minimum amounts of hazardous pesticides. Such a policy would, moreover, further economic goals. Production setbacks due to pest or disease problems will undermine development and threaten the financial viability of a project. The problems will not, in the long term, be solved simply by the use of pesticides and disease resistant varieties of crops. Their solution can best be achieved by anticipating pest and disease problems at the outset and devising research and management techniques accordingly.

The management technique that minimizes the amounts of toxic or environmentally harmful pesticides employed is 'integrated pest control'. It is one that also reduces the cost of control over the long term, and the potential for disruptions in insect pest populations that can lead to uncontrollable outbreaks. Integrated pest control requires a combination of investigation and technology, and therefore concerns both project planning and management.

Integrated control has been defined by Smallman (1965) as 'The deliberate, artificial mimicking of just enough of that diversity of selection pressures seen acting in nature, to relieve the selection pressure from any one agent, and thus prevent the pest getting out of control by developing resistance faster than we can revise our control measures.' Its general aim is to use a mixture of biological, cultivation and chemical means to keep pest populations at levels that are not economically significant.

This management technique was employed by cotton producers in the Canete Valley of Peru after excessive use of organochlorine and organophosphorus pesticides had disrupted insect ecology so drastically that production had become uneconomic. The integrated programme included curtailment of cotton production on marginal lands, prohibition of ratoon cotton, the introduction from neighbouring valleys of a new stock of the natural enemies of cotton pests, synchronization of cultivation time and methods, a cotton-free fallow period, and return to nicotine sulphate and arsenical insecticides (the latter, however, having the disadvantage of being relatively persistent). These measures, which were codified as regulations. were approved by all growers. Similar integrated control of cotton pests has been recommended in Guatemala and is already being practised in Mexico.

Integrated pest control is most effectively achieved by viewing the agricultural landscape as an ecosystem. A description of the agroecosystem, based mainly on Smith and van den Bosch (1967) and Southwood and Way (1970), is therefore appropriate at this point.

An agroecosystem consists of the total complex of organisms in the crop or a cultivated region, as modified by the various agricultural, industrial and other activities of man. Its major components include crop plants, the soil and its essential biota, the physical and chemical (natural and artificial) environment, energy from the sun, and man. Temporary elements such as weed species, plant disease pathogens or insects may become critical or dominant elements in the system. For the development planner, the boundaries of an agroecosystem could be the limits of an irrigation district, a watershed, or the supply region of a major marketing centre. For an entomologist they might be delimited by the spatial distribution and habitat of a population of insects, with their predators and parasites. Boundaries would be defined according to the particular management problem and the geographic extent of the system to which it pertains.

Agroecosystems may have stable vegetation, such as coffee, or a discontinuous cover of annual crops, punctuated by periods of bare soil. They may contain a complex of different species, both perennial and annual, but generally not very diverse. One or two species tend to be dominant at a time, and to be at a uniform stage of development. Compared with most natural ecosystems, agroecosystems have greater amounts of nutrients in their foliage and more young growing tissue. Pest and disease outbreaks are regular features and pests tend to adapt themselves readily to changing or temporary agricultural environments. These explosive outbreaks signify agroecosystem instability, associated with lack of species diversity and relatively few connections or trophic interactions between plants and insect species. An ecologically-oriented agroecosystem management plan would aim at keeping insect populations relatively stable and at tolerable levels, through the maintenance of crop and vegetation diversity or the management of crops, boundary vegetation and insect predators so as to encourage many interactions between species.

The diversity or simplicity of plant species in an agroecosystem has a varying relationship to that of the natural ecosystem. Thus, monoculture in the tropics is a greater departure from natural species diversity than monoculture in colder latitudes. At the same time, the natural ecosystem of the tropical humid forest is one of the most stable in the world, by virtue of its great diversity, and its climate is equally stable, compared to one in which rainfall is markedly seasonal.

Major objectives of integrated pest control are to avoid the development of resistance to insecticides and to minimize disruptions to the ecology of predators and parasites which prey on the insect pest. In Table 3 the development of resistance of some major agricultural insect pests is summarized. As will be seen, the time interval between first use of the pesticide and the development of resistance in these pests ranges from two to seven years.

The rapidity with which resistance develops is dependent upon such factors as the rate of application of the pesticide and the time elapsed between generations of insects. However, it is difficult to predict the emergence of pesticide resistance due to the complex ecology of the insect

Table 3. Some examples of the development of pest resistance to pesticides (after Brown, A. W. A., 1957)

Insect	Pesticide	Time lapsed from start of application to observed resistance	Year resistance observed	Crop	Location
Citrus thrips (*Scirthothrips citri*)	Tartar emetic	2 years	1945	Citrus	San Fernando Valley, Calif.
Coffee thrips (*Diarthrothrips coffea*)	DDT	no data	1956	Coffee	Tanganyika
Lygus hesperus	DDT	7 years	1953	Alfalfa	Yakima Valley, Wash.
Grape leafhopper (*Erythroneura variablis*)	DDT	6 years	1951	Grapes	Coachella Valley, Calif., Phoenix, Ariz.
Grape leafhopper (*E. elegantula*)	DDT	no data	1951	Grapes	Tulare County, Calif.
Apple leafhopper (*E. lawsoniana*)	DDT	no data	1953	Apples	Paducah, Ky.
Walnut aphid (*Chromaphis jugandicola*)	Parathion	no data	1953	Walnuts	California
Cotton aphid (*Aphis gossypii*)	Lindane (gamma-BHC)	no data	1953–7	Cotton	Southeastern USA
Cotton leafworm (*Alabama argillacea*)	Toxaphene	5 years	1951	Cotton	Texas
Spring bollworm (*Eariàs insulana*)	Endrin	3 years	1956	Cotton	Israel
Diamond-back moth (*Plutella maculipennis*)	DDT	3 years	1951	Cabbage	Java
Boll weevil (*Anthonomus grandis*)	Toxaphene, dieldrin, endrin and lindane	no data	1954–5	Cotton	Mississippi delta, Red River Valley

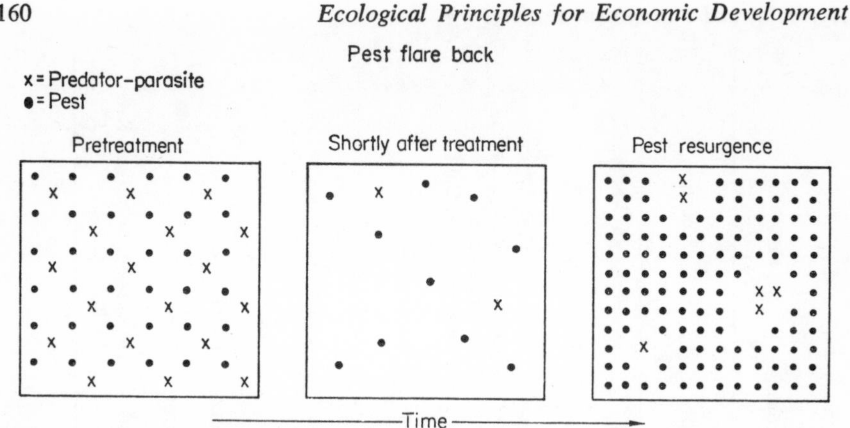

Figure 22. Diagrammatic sketch of the influence of a chemical treatment on natural enemy-pest dispersion and resulting pest resurgence. The squares represent a field or orchard immediately before, immediately after, and some time after treatment with an insecticide for control of a pest species represented by the solid dots. The immediate effect of treatment is a strong reduction of the pest, but an even greater destruction of its natural enemy (enemies), represented by × 's. The resulting unfavourable ratio and dispersion of hosts (pest individuals) to natural enemies permits a rapid resurgence of the former to damaging abundance. (Smith and van den Bosch, 1967).

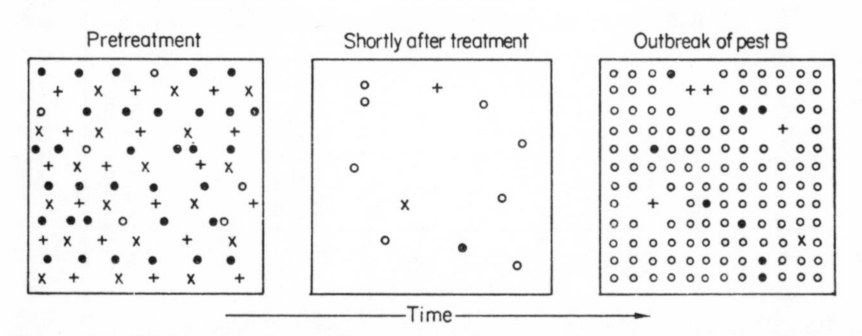

Figure 23. Diagrammatic sketch of the influence of a chemical treatment on natural enemy-pest dispersion and the resulting secondary pest outbreak. The squares represent a field or orchard immediately before, immediately after, and some time after treatment with an insecticide for control of pest A, represented by (●). The chemical treatment effectively reduces pest A as well as its natural enemy (×), but has little or no effect on pest B (o) present in low numbers before treatment, but devastates the natural enemies (+) of pest B. Subsequently because of its release from predation, pest B flares to damaging abundance. (Smith and van den Bosch, 1967).

populations and impact of the pesticide on this ecology. Figures 22 and 23, taken from Smith and van den Bosch (1967, pp. 298 and 299) illustrate the disruptions in insect populations which can be caused by pesticides. If the pesticide kills the pest predator as well as suppressing the pest population, the latter may flare back in unprecedented numbers (Figure 22), or a new pest may emerge due to the inadvertent suppression of natural enemies which held it at non-destructive levels (Figure 23).

These figures also serve to illustrate the probable presence in an agroecosystem of a great number of insects which are not destructive of crops but may have a parasitical or predatory relationship with crop pests, or indirectly contribute to the stability of the insect population. In Southern Californian alfalfa fields, one thousands species of arthropods were estimated to be present or associated with this crop. Certain irrigated cotton fields in that state contained an estimated 300 to 350 arthropod species, only 20 per cent of which were phytophagous, i.e. feeding on plant matter (Smith and van den Bosch, 1967, p. 306). Thus, the insect life of a seemingly simplified agroecosystem may be exceedingly diverse, a circumstance which both contributes to the stability of insect populations and also makes integrated control a complex problem.

The firm prediction of insect population instability and destructive pest outbreaks is not possible without detailed knowledge of the response of insects, both pests and their predators, to changes in food supply. Important characteristics in the agroecosystem which influence this response are diversity of the vegetation, permanence of the crop, stability of the climate and degree of isolation. A case of extreme isolation would be a desert oasis or irrigated cultivation surrounded by desert. An even-aged and genetically uniform crop provides exceptionally favourable conditions for a rapid increase in pest population, especially if its growth is synchronous with that of the pest. A deliberate policy of promoting instability of plant cover, in order to deprive the insect pest of off-season host vegetation, may be adopted and effective but is not applicable in the humid tropics, where the goal of attaining maximum productive potential hinges on continuous cropping.

By contrast, the pattern of spread of a disease in a crop after its first appearance is usually somewhat easier to forecast. This generalization applies mainly to fungus type diseases, such as *Helminthosporium* spp. (leaf blights), whose spores are carried by wind and rain, rather than to virus diseases. The latter may be spread by insects, which, for the purposes of prediction, entails the additional consideration of a set of variables related to the insect agent's ecology and greatly complicates the construction of a predictive model.

To sum up, there are several basic assumptions which can be made at the project planning stage of integrated pest control:

(i) *An extensive area of a genetically uniform crop represents the agroecosystem which is most vulnerable to insect and disease outbreaks.*

(ii) *The greater the departure from natural ecosystem diversity to agroecosystem simplicity, the greater is the potential damage from pests and diseases.*

(iii) *Since the more intensive the use of insecticides, the greater the pressure for selective resistance, the result of such use may well be insect population instability and uncontrollable outbreaks of pests.*

While these assumptions are subject to qualification, they can be generally useful when the scope and production goals of an agricultural development project are being considered. They also carry implications relative to research and plant breeding facilities, technical, advisory and credit services, and storage facilities—investment in which is recognized as essential for the support of any farming system which relies on modern technology or, in other words, for the maintenance of productivity in all high-yield agroecosystems.

Experience of insect pest control has shown that an integrated approach, including the judicious and selective use of toxic chemicals, is the only suitable one, if pests are to be kept at levels at which they do not constitute a serious economic problem. Integrated pest control programmes are also the most promising answer to the environmental hazards of toxic chemicals and the predictable insect population instability resulting from their excessive use. To the extent that such programmes involve intercropping, the types, spacing and timing of crops, and other crop management features, they will have a direct effect on production goals of any agricultural development project. It is consequently of importance that, at an early stage in project planning, the advice of entomologists should be sought, to ensure that objectives are compatible with the ecological dictates of integrated control.

At the project planning stage, therefore, preinvestment studies should include a compilation of relevant experience and direct field observations to establish, to the extent possible, the economic injury levels of pests liable to affect production, and to determine their ecology. The results should be incorporated in the crop and soil management plans for the project, as these may be affected by integrated pest control measures. The continuing research and technical advisory services needed for integrated control programmes should also be defined in relation to their manpower, budget and administrative requirements.

When a project is launched, the management of integrated pest control should be an explicit objective and cover the continuing research on pest ecology, economic injury levels and pest response to integrated control, and the technical advisory services to coordinate the programme. To the extent that the use of potentially toxic or environmentally hazardous pesticides is inevitable, monitoring of pesticides levels in crop, soils, water and non-target organisms should be initiated.

Some countries may not have sufficient expertise and institutional capabilities to carry out these activities. Their manpower and institutional needs should be defined as early as posisble in the project planning stage, and the necessary training and advisory services determined. Long term research and technical assistance such as that offered by FAO may be required.

(g) Ecological Aspects of Weeds and Weed Control

For the farmer, weeds are plant species that grow where they are not wanted, competing with crops for moisture, sunlight and nutrients. They have been called plants that are out of place (Ochse *et al.*, 1961). However, from an ecological standpoint weeds very definitely do have a place in an ecosystem, and the magnitude of the problem they represent to the farmer is a measure of the degree to which agricultural land use creates opportunities, or niches, for their spread. Weed species tend to be vigorous and persistent invaders of exposed soil in cultivated fields and plantations. Such unutilized or uncolonized sites rarely exist in natural plant communities. Weeds are well adapted to temperature extremes in bare soil of planted fields prior to the emergence of the crop, and the tolerance of some species to low nutrient status is greater than that of crops.

In the humid tropics and subtropics, plots cultivated by the subsistence farmer have, as a rule, been abandoned after several harvests have been taken, although the soil may not be totally depleted of nutrients. The practice ensures that robust and aggressive weed species will not dominate or delay forest regrowth—an important consideration for the farmer who depends upon a 'good quality' forest fallow for the nutrients that will be released on next felling and burning. The use of chemical fertilizers obviates the need for this traditional technique and makes continuous cropping feasible. The ability of a farmer to control weeds may, however, constitute a limiting factor in the amount of land he can cultivate. While mechanized or draft animal power may enable him to plant a larger area and fertilizers to crop continuously, the labour required for hand control of weeds may be a serious obstacle. Herbicides are, therefore, viewed as virtually indispensable for intensive, high-productivity farming in the tropics, especially in the case of short-cycle crops.

Herbicides are, fortunately, not as persistent as organochlorine insecticides.

With a few exceptions, such as sodium arsenite, they are toxic only to plants.* Of necessity their phytotoxicity is selective. Most herbicides are not water soluble, and thus do not contaminate runoff, but remain in the soil. This circumstance, however, compels a careful evaluation of all their possible effects on plant life during the time of persistence and especially of their impact on rotation crops and mixed cropping schemes.

Although not as great a threat to the environment as some insecticides, herbicides are not totally lacking in environmental side effects that could be unwanted. Some of the toxic and ecological effects of 2, 4, 5-T are an example. An apparently unavoidable impurity introduced in minute amounts during manfacture of this herbicide, is dioxin, which is not only highly toxic in small doses but in experiments has produced birth defects in mice. 2, 4, 5-T has been implicated in as yet unsubstantiated cases of human birth defects in Vietnam, where it has been sprayed for military purposes, although there is currently major disagreement among scientists as to whether or not the use of 2, 4, 5-T exposes humans to hazardous amounts of dioxin. However, other undesirable effects have been more certainly observed. 2, 4, 5-T has been reported to reduce egg production in domestic fowls and to be quite toxic to certain fish and oysters. In some plants treated with the herbicide, nitrate content has been increased to levels more toxic to mammals. Similarly, hydrocyanic acid content of sudan grass was found to increase by 70 per cent following 2, 4, 5-T applications of 1 lb/acre (Panel on Herbicides, 1971). Finally, although this herbicide apparently does not persist beyond three months in soils, mangroves killed in Vietnam with aerial sprays of 2, 4, 5-T and 2, 4-D failed to regenerate after six years.

Other complications pointing to great caution in herbicide use for weed control have to do with uncertain, and possibly synergistic, effects of interaction between different herbicides, and of herbicides with pesticides, fertilizers and crop plants. The full extent of these interactions have not been identified and evaluated, but could be unexpected. Even the exact nature of the toxic reaction to plants is not yet understood.

For instance, it was observed that simazine (of the triazine group) applied around fruit trees enhances shoot growth, and later experiments showed that this herbicide applied at very low levels substantially increased the protein in rye, peas and beans, especially on nitrogen deficient soils (Wittwer, 1970). On the other hand, a deleterious result of the use of dicamba on cereal crops in Britain which had not been anticipated, was subsequent injury to other susceptible crops when the residue-contaminated straw was used for mulch (Official Report, 1970).

The report just quoted, presented by the U.S. delegation to the First

* Of the organic herbicides, the two most poisonous to mammals are dinoseb-LD_{50}, 5 to 60 mg/kg; and endothall-LD_{50}, 38 to 51 mg/kg.

International Conference on Weed Control, also included the following cautionary statement (Official Report, 1970, p. 32):

'Better analytical methods are required, and more research is needed on soil analysis, half-life studies, and residue determinations in plants, soils, water, fish and foods. The consequences of contaminants and formulation ingredients in herbicides should be studied. Users of herbicides urgently need better guidance and means of disposal of excess herbicides and containers. Misuse of herbicides could often be avoided by more complete and effective instructions, and if existing direction and guidelines were followed.'

The control of weeds presents several parallels to the ecological problems associated with the control of insects: shifts in weed populations to species that are resistant to the chemical used, and great vigour resulting from an abundance of nutrients (fertilized soils in the case of weeds; the crop itself in the case of insects). In short-cycle, or annual crops, weed competition is most serious at early stages of plant development, when weed species grow faster than the crop. At later stages, when the crop is developed, it competes more successfully with weeds for sunlight and nutrients. It is notable that the short, stiff straw characteristic of high-yield wheats has diminished their capability to compete and a further disadvantage is posed by the upright leaves which also characterize these varieties: more sunlight reaches the soil, which has received abundant nitrogen fertilizer, and conditions for weed growth are therefore very favourable. So the cultivation of these high-yield wheats becomes impossible without the use of herbicides or extensive hand weeding.

In tropical plantations of coffee, cacao, sugar cane, oil and coconut palms, and other crops, losses in yield due to weed competition have not been accurately determined. However, periodic weed-cutting, usually with machetes, in tree crop plantations is common practice. In regions of abundant moisture, uprooted weeds must be physically removed since they will take root and survive. Tillage of the soil does not control perennial weeds, and may actually stimulate weed growth. Hence, herbicides are regarded as a good control for weeds around the trees in plantations. Nevertheless, certain herbicides may be toxic to young coconut palms and oil palms and are therefore avoided, while the herbicide bromacil has been found to be toxic to mature coconut palms (Hoyle, 1969). Some other oil-seed crops are also adversely affected by particular herbicides (Official Report, 1970).

Although of relatively short persistence, herbicide residues in the soil may be toxic for other crops. Sugar beets are injured by residues of trifluralin and atrazine. Atrazine is especially persistent and toxic to such crops as small grains and soy beans, which may be grown in rotation with corn (maize) treated with atrazine.

Such practices as interplanting corn and beans, and phased planting, reduce opportunities for weeds to become established. The employment of a useful weed-exclusive cover crop, such as a pasture grass in coconut and rubber plantations, also provides control and, in addition to protecting the soil and maintaining organic matter, supplements land use, in the example mentioned by supplying grazing. The economics of such multiple land use have to be weighed against those of a monoculture in which herbicides may have to be utilized before their economic effectiveness can be definitely established. In general, it seems a reasonable assumption that the return from a suitable cover crop will compensate for any lower yield of the primary crop due to competition for sunlight and nutrients.

From the strict standpoint of weed control, the proximity to cultivation of zones of disturbed secondary vegetation, containing vigorous colonizing species, is often considered undesirable, because such zones provide a ready source of weeds. They may also contain off-season host plants for insect pests and disease pathogens (Smith and van den Bosch, 1967, p. 305). The eradication of vegetation in these contiguous zones is therefore sometimes advocated, but for ecological reasons may be unwise. They may well provide a habitat for the insect, bird and mammal species which prey on plant pests and help to maintain these at acceptable levels. Use of herbicides sufficiently toxic to eradicate all possible weed species from boundary zones could also result in death of other more favoured plants. Furthermore, the probability is that, following herbicide treatment, a few extremely hardy species, capable of becoming serious weeds, especially in the tropics, will become established. In Vietnam, for example, forests killed by aerial sprays of 2, 4-D and 2, 4, 5-T have been colonized by cogon grass (*Imperata* sp.), and scrub bamboo, both extremely difficult to eradicate, of no economic importance and so vigorous that broadleaf tree species are unable to re-establish themselves (Westing, 1971).

For all these reasons, it is clear that the use of herbicides for weed control needs to be under close and competent technical supervision. It should moreover be closely coordinated with or part of an integrated pest control programme, since weeds may constitute food or habitat not only for insect pests but also for their predators and parasites. Agricultural development projects should therefore include technical assistance and research in weed control; and manpower, supporting services and budgetary requirements should be provided accordingly in the project management.

(h) Ecological Aspects of Disease-resistant Plants and High-yield Varieties

Development of disease-resistant varieties of crop plants is normally accomplished by cross-breeding with existing cultivated varieties known to be resistant to a particular disease. There are several ecological complications associated with these techniques for combating disease. Varietal resistance does not remain commercially effective because of the versatility and diversity of pathogens attacking crops, and the life expectancy of most resistant cultivars and hybrids is likely to be 5 to 15 years (McNew, 1966). Secondly, varietal resistance is unlikely to be universally effective due to geographic variations in plant pathogens. A further complication arises from the disappearance of genetic types of the crop plant, which although resistant do not produce high yields or are of poor quality. These varieties, which could once be found quite frequently in subsistence or old-fashioned farming areas, are being rapidly displaced by higher yielding varieties.

High-yield varieties, such as the Mexican miracle wheat and the several varieties of rice bred at the International Rice Research Institute in the Philippines, were of course bred for disease resistance, but are subject to the complications mentioned above. In addition, because of their very success they present a rich and attractive source of food for insects and other animals. Since these high-yielding and disease-resistant strains form the basis of the 'Green Revolution', careful notice of associated ecological complications is merited (see Stakman, E. C. (1967), Brown, L. R. (1970) and the annual reports of the Rockefeller Foundation).

(i) Impermanency of Disease Resistance

The attainment of resistance to virus, fungal or bacterial diseases is a major aim of plant breeding programmes. In the advanced agriculture of the USA, plant disease continues to cause heavy losses, estimated at $3 billion yearly, notwithstanding considerable research and experimentation since the early part of this century. Disease-producing agents, particularly viruses, seem to have originated locally throughout the world, and through long association with the pathogens, native plants have developed a measure of immunity or tolerance to them, and are able to survive with little injury and often with little evidence that they may be carriers of disease. Thus, wherever a serious disease occurs in nature, plants capable of resistance may be found, because otherwise there would have been no survival. It is the disease-resistant traits that the plant breeder searches out and introduces into botanically related economic plants. However, plant breeders recognize that it is a never ending process.

As McNew (1966) remarks—'Man has never controlled plant diseases. We have merely learned to live with them. There are probably no more than six diseases that have been totally suppressed, much less eradicated. . . . The perfect disease-resistant crop is almost a dream because of the versatility and diversity of pathogens attacking crops.' He goes on to say that while it may take good scientists ten years to work with an easily handled crop such as cereals or potatoes, to locate the source of resistance, transfer the genes into commercially acceptable varieties and, after back-crossing, multiply the stock sufficiently for commercial usage, the pathogen may produce a new race in ten days, and build up a reservoir of inoculum in less than one season. Pathogens have their own breeding programme.

A prime requisite for a plant breeding programme whose goal is to develop disease resistance is to select and breed the resistant traits in conditions that parallel those of natural environmental conditions. Unless the exposure to disease encompasses the ordinary range of virulence exhibited by the gamut of biotypes of the pathogen likely to be encountered, the resistant strain developed may be very restricted as to its area of adaptation and short lived as an agricultural variety. (Coons, 1953; McNew, 1966).

(j) Disappearing Genetic Resources

The success of agricultural development and of the Green Revolution in particular—measured in terms of rapid adoption of high-yielding varieties—threatens the existence of hardy native species that could prove crucial to the maintenance of the high-yielding plants. To understand this, it is necessary to consider in some detail the methods by which the new varieties have been produced.

High-yielding varieties of wheat and rice show a response to large amounts of fertilizer, especially nitrogen, which is expressed in very large heads of grain. Another essential characteristic of these varieties, therefore, is the short, stiff stalk needed to support the heavier load of grain. Few of the thousands of varieties of rice and wheat, or their progenitors, possess the requisite character of a short, stiff stalk. In the case of wheat, it was derived from a dwarf variety, representing a combination of Japanese and other strains and adapted to the Pacific north-west of the USA, where it is known as Gaines wheat. A dwarf rice from Taiwan provided the short stalk characteristic for the high-yielding rice, IR-8.

Some of the high-yielding cereal varieties are already proving susceptible to disease, and our knowledge of the versatility of plant disease pathogens gives cause to suspect that any resistance they presently enjoy will eventually collapse. As Borlaug, quoted by Stakman (1967), has emphasized, the achievement of high-yielding varieties must accordingly be complemented by continuing breeding programmes in each country to develop and maintain disease resistance. Very local varieties of wheat and rice, and their wild progenitors, may be the key to accomplishing this. These native plants have been exposed to the local diseases for many years, or even centuries in the case of certain wheat in Asia Minor. Although they may not be altogether immune, they may have genetic traits that could contribute to resistance to prevalent diseases and, incidentally, to better adaptation to local growing conditions, which could usefully be transferred to new high-yielding varieties.

Unfortunately, as has been pointed out in an interview with four leading plant geneticists, published in *Ceres* (Round Table, 1969), there has recently been a rapid erosion of these genetic resources. In Africa, one of the centres of origin for rice, less than 10 per cent of the rice grown in the plain of the Upper Volta is now of African origin, although in 1935 it amounted to 85 per cent. It has been supplanted by Asian rice. The disappearance of the local rices would be an irretrievable loss. Total losses of plant materials have already been recorded in the Cilician plain of Turkey, the geographical origin of diverse varieties of flax. Not one local variety survives and the varieties found in Argentina are all of cultivated origin. Thousands of varieties of *Triticum spelta*, a relative of wheat, once existed in Iran, but have disappeared completely in the last 15 years. Native varieties of wheat are rapidly disappearing from Asia Minor and also from Ethiopia, where they were formerly very numerous. In Turkey, the wild progenitors of wheat and other cereals are found only in graveyards and the ruins of castles where they have protection from grazing animals. The collection of native maize varieties in South America, undertaken in the 1960s, was none too soon. Improved varieties and hybrids had already begun to replace the ancient maizes of Colombia, some of which are now hard to find (Stakman, 1967).

The significance of disappearing genetic resources can be appreciated by hypothesizing the loss of disease resistance of a major food grain. The possible impact of a major outbreak in plant disease on rice is illustrated in Figure 24. This hypothetical example is patterned after data on increase in rice production from the expansion of area under high-yield varieties in the Philippines (Keefer, 1969).

The realistic assumption is made in Figure 24 that disease resistance will collapse from five to fifteen years after the resistant variety has been developed. The collapse of resistance to such diseases as brown spot, narrow brown leaf spot or blast is estimated to occur in the hypothetical example at about the time that extensive and contiguous irrigated areas are under the production of a single high-yield variety, such as IR-8, at which time the diffusion of this variety peaks out (1974). The diffusion rate of the pathogen is here assumed to be rather rapid. Its emergence stimulates efforts to breed varieties to be selected for resistance, which by 1976 have produced several that are under trial. In 1977, an effectively resistant variety is offered to rice growers, and by 1981 has replaced the susceptible variety. In this example, total production is inferior to demand during four years. Because the rapid spread of disease could depress production below the prospective food needs of a society, it is clear that

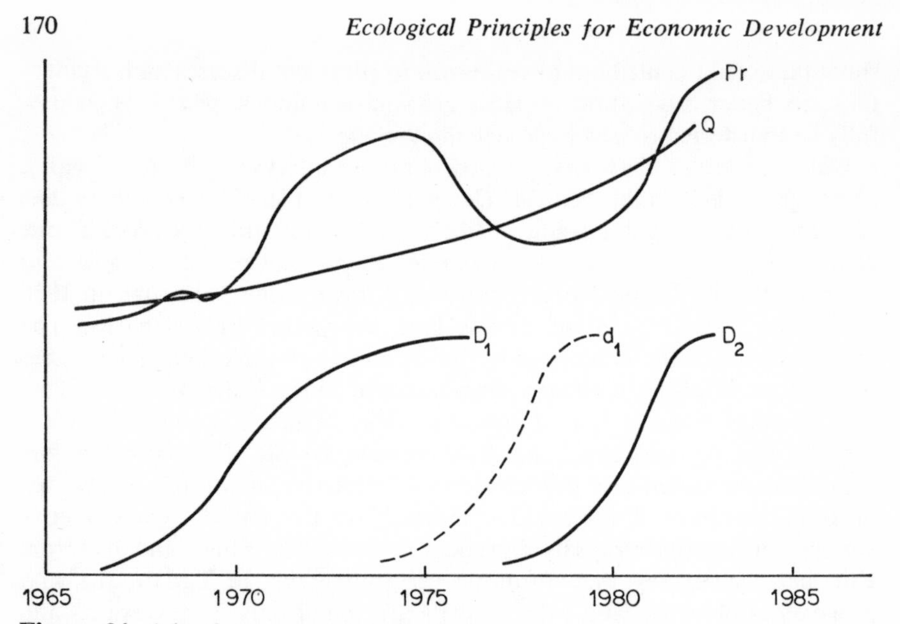

Figure 24. Adoption rates of high-yield varieties, diffusion rate of disease pathogen, internal demand and total production.

- D_1 Diffusion curve of high-yield grain variety
- d_1 Diffusion curve of disease pathogen
- D_2 Diffusion curve of new, disease resistant high-yield variety
- Q Total internal demand curve, in terms of food needs
- Pr Total production curve.

not only must there exist plant breeding capabilities for quickly developing resistance to new races of pathogens, but that large stocks of grain must be stored in anticipation of temporary shortages. Ultimately, success in outpacing new strains of disease hinges on the preservation of genetic resources.

Threatened genetic resources can be and are being salvaged by maintaining plant collections and by placing seeds in cold storage, serving as gene banks. These solutions are better than none, but are imperfect to the extent that natural evolutionary processes and the process of natural selection for local environmental conditions, including disease, are interrupted. An alternative or additional safeguard would be to maintain, in representative ecosystems, areas of native vegetation uncontaminated by new and exotic varieties.

A major threat to genetic resources for future breeding programmes is posed by the opening-up of previously inaccessible regions by roads. With the consequent shift from subsistence to commercial farming, food plants which, over centuries, have been selected for their suitability in local ecological conditions, are supplanted by a small range of marketable varieties, very probably of external origin. But because the local varieties thus threatened with extinction may well be a valuable future source of genetic material, it is imperative that they should be collected and maintained in botanical gardens or other suitable depositories, *before* the major land use changes become operative. The same applies of course to the collection of potentially important material in areas due to be inundated by the construction of major dams.

(3) SUBSISTENCE FARMING—ECOLOGICAL AND RELATED PROBLEMS

The problem of increasing productivity on the small farm is especially urgent in developing countries characterized by high rates of underemployment or unemployment in both rural and urban areas, and by an unmanageable influx of rural migrants to cities. A further and most serious aspect of the problem is deterioration of the environment in consequence of excessive demographic pressures on land resources. The major policy questions involved can be answered in part by determining the degree to which land resources have been saturated by subsistence farming populations. It has been estimated that subsistence farms cover some 40 per cent of the cultivated land of the world and support 50–60 per cent of mankind (Wharton, 1970). Approximately a fifth of the farmers obtain the bulk of their food by the system of shifting cultivation (Nye and Greenland, 1960); the remainder are relatively sedentary.

(a) Carrying Capacity and the Subsistence Farm

The amount of land needed to sustain a family in the tropics and subtropics shows considerable variability, from 50 or more hectares per shifting cultivator farming rapidly depleted soils in humid lowland forest, to one or two hectares or less of fertile and well-watered alluvial soil, capable of continuous cropping. Carrying capacity in this context is related to the number of people whose food needs can be satisfied by production from lands under traditional food crops, at land use intensities which do not destroy the resource base.

Determination of carrying capacity for subsistence farming is complicated by the fact that few subsistence groups have no contact whatsoever with markets, and in fact some cultivate non-food crops, such as kenaf, rubber, oil palms or tea, which are traded for food items. The subsistence level carrying capacity of a region is, therefore, more accurately, a measure of the amount of food that can either be produced or obtained in exchange for production. A gradient could be devised showing subsistence farms according to the proportion of food actually produced on the farm—from totally self-supporting units to farms that exchange all of their production for food. The latter extreme is unlikely; there will usually be at least some small garden plots devoted to staples such as manioc, corn or rice. However, where a considerable proportion of food is obtained outside the farm, the measure of carrying capacity is modified by external markets for commodities traded and the availability of imported food. In other words, environmental factors influencing the production of cash crops are not accurate indicators of how many people or how many small farms a region can accommodate, or how large a farm must be to provide at least a subsistence diet.

The small farmer who trades most of his production for food is especially vulnerable to market changes. A minor price change in either the commodity traded or the food item bought may mean a restricted diet and malnutrition. It could be hypothesized that if the population enjoys a good diet—that is, does not suffer from seasonal weight loss or nutritional diseases—the carrying capacity of a region has not been surpassed. However, it is possible in this context that, even though the subsistence carrying capacity of a region may not have been surpassed by population, people are inadequately fed. This could come about for several reasons: a fall in market prices of the commodity traded relative to food prices; food purchases inadequate from a nutritional standpoint; or climatic adversities resulting in an absolute deficit of food in the larger society and a curtailment of imports.

An increase in both efficiency of production and total production of a commodity may seriously affect the small farmer who operates at the subsistence level and depends on a small income from producing the commodity. A major innovation, such as a high-yield grain, can substantially alter the supply situation and depress market prices to the point where larger farms with lower production costs dominate the market and smaller producers are unable to compete. This has already been observed in India in relation to high-yielding wheat varieties (Shaw, 1971). Small farmers located far from, as well as near to, a project that brings about production increases may be affected.

Since economic development is bound to involve farmers in the market system, however it may be structured, it is of utmost importance to consider how a continuous and adequate supply of food will be provided to the subsistence farmer, in those situations where the aim of development is to stimulate the production of cash crops in place of subsistence food crops. If the detrimental effect of market fluctuations on food supply are not countered, the reaction of the small subsistence farmer will either be to switch from the cash crop back to subsistence food crops or to exploit his land more intensively, in order to compensate for reduced food purchasing power. Either alternative could be destructive to the environment. In view of these factors, it would be overly simplistic to attribute environmental deterioration associated with subsistence farming to the condition of overpopulation only. For many countries where small farms attempt to produce cash crops but use the income for food items, the ability of the food distribution and marketing system to adjust fluctuations in food prices and supplies so that subsistence-level populations are not adversely affected, is a key problem of possibly just as great significance.

It is quite obvious, however, that some regions can support denser populations than others, for instance fertile alluvial plains as opposed to rocky uplands. Even though fertile areas may be dominated by small farms with low standards of living, they have usually been settled for a long time and have a fairly well developed infrastructure and food distribution system, with local markets where produce is traded. Marginal environments (typically devoid of infrastructure) which are only recently settled or whose formerly sparse populations have exceeded the subsistence carrying capacity, are those which are likely to be deteriorating as a consequence of overintensive or destructive uses (Figure 25). Environmental deterioration is most dramatically seen in sub-humid and semi-arid tropical environments populated by subsistence farmers or pastoralists, and especially in terrain with steep relief. Ecosystems with these climatic

Figure 25. Subsistence farming on marginal land in Kenya. Low yields and accelerating land damage impoverish people and environments together. (Photograph by Kenya Information Service).

regimes are less resilient than the humid tropical ecosystem, where vegetative growth is vigorous and where many food and cash crops are perennial, and thus protective of the soil resource. It is in these drier regimes that the process of desertification described in Chapter 4 is evident, and that short-cycle food crops which expose the soil to erosion are typically grown. Charcoal-burning and goat-herding are also common activities and can contribute to the steady degradation of vegetation. The other main feature accounting for the fragility of these ecosystems is the variable nature of rainfall; droughts may aggravate the deterioration initiated by browsing, overgrazing and charcoal-making, while unusually heavy rains can cause catastrophic erosion on slopes cleared for food crops.

An illustration of a region where subsistence level carrying capacity has apparently been exceeded by population is provided by the Western Altiplano of Guatemala (CIDA, 1965). This mountainous zone, 1,500 to 3,000 metres in elevation, contains the majority of the 350,000 farms in Guatemala which, in 1950, were considered too small to support the families which live on them. Rainfall is seasonal and allows only one crop per year. Farms range from less than one-half of a hectare to two hectares in size and are owned and operated mainly by Indian descendants of the original pre-Colombian population. Most of the region dominated by these small farms is suitable only for forestry, and those lands with potential for agricultural production have been judged to be only moderately productive, even under intensive production practices (Plath, 1968). In a sample of small subsistence farms studied in this region, it was found that 90 per cent of each farm was cultivated (i.e. little or no fallow) for the production principally of maize and, to a lesser extent, beans. Maize is the traditional crop and, in the absence of sufficient land to produce other products, provides the greatest nutritional value. Yields were 800 kg/ha, in comparison with 1,200 to 2,250 kg/ha from better lands in the coastal lowlands cultivated by small farmers. Cultivation activities occupied only 110 man days per year, and the need to seek other means of subsistence resulted in mass seasonal migrations of an estimated 160,000 persons (in 1963), to work on coffee plantations during harvest time. An increasing amount of steeply sloping, poorer quality land was being cleared, and soil erosion was observed to be serious. 50 per cent of any additional income the farmers earned was for food, mainly meat, coffee, salt, sugar, chilis and rice. Infant mortality in one area accounted for more than half the recorded deaths. In another region 11 per cent of the deaths were judged to be from malnutrition. Yet the indigenous farmers were loath to give up their land, and properties were increasingly fragmented and divided as families grew.

Similar examples could be found, with slight variations, throughout the highlands of southern Mexico, central America and the Andes; and again in Africa, India, Indochina, the Philippines, Indonesia and elsewhere in tropical and subtropical countries.

In the more humid tropics, overpopulation relative to carrying capacity is more likely to be evidenced by malnutrition, especially animal protein deficiency (kwashiorkor), and environmental deterioration may not be so dramatic under subsistence methods of resource exploitation. On the other hand, as pointed out in Chapter 3, it will very probably be modern methods of agricultural and forestry exploitation that are doing the more serious damage.

To sum up, the significance of high population densities in subsistence farming communities must be measured in terms of:

(i) the biological productivity of the environment in relation to the production of food and other useful products;
(ii) susceptibility of the environment to deterioration under subsistence cultivation practices;
(iii) the nutritional value of food consumed, whether actually produced or obtained in exchange for production; and
(iv) the dependence of the subsistence farmer on external economic factors over which he has little control and which affect his ability to obtain adequate food, such as commodity markets and food distribution systems.

The environmental consequences of destructive land uses associated with excessive densities of subsistence farmers vary considerably according to ecosystem characteristics and type of subsistence exploitation. Remedial measures must take into account environmental and human ecological problems, the solution of which may entail not only major changes in land use but also the provision of infrastructure and the control of market mechanisms affecting small farmers.

(b) Implications for Agricultural Development Planning

This brief diagnosis of ecological aspects of subsistence farming illustrates the complex nature of the problems facing the development planner. The nutritional and food supply problem is, clearly, of paramount importance, but it is seldom if ever included as a development objective in production or resource conservation programmes affecting small farms. It is a human ecological problem which is not necessarily solved by commercializing subsistence farming and obviously cannot be diagnosed only in terms of commodity production, income or price statistics. It needs to be approached from the standpoint of the subsistence farmer's own basic objective; to feed himself and his family. Any development project which does not fulfil this fundamental objective is bound to fail.

Adjustments in land use following technical innovations are to be expected. However, the proposition that more efficient agricultural production will stimulate industrial growth and generate employment in cities begs the question of what is to become in the interim of the disadvantaged small farmer and the environment that he occupies. The fact remains that environmental degradation due to population densities in excess of carrying capacity is already in evidence in many parts of the tropics, particularly in sub-humid regions with a hilly terrain.

If projects aimed at increasing the production of subsistence farmers are to be ecologically sound as well as economically feasible, preinvestment studies should include, in addition to other standard physical and socio-economic data, a detailed analysis of all aspects of food production,* an estimate of carrying capacity of the project area at existing and contemplated levels of technology, and an assessment of land tenure with special attention to characteristics related to traditional food production methods.

Projects which could adversely affect the small farmer's share of the market should provide for whatever remedial measures may be needed to guarantee minimum food requirements and assist him to adjust his operations so as not to lose income because of more efficient competition. Special attention should be paid to farms located in densely populated zones, whose resource base may rapidly deteriorate under intensified subsistence farming. The market region served by the new production would determine the location of potentially affected small farm units. Preinvestment studies should include a survey of those farms which would be put at a competitive disadvantage, a determination of carrying capacity of the areas they occupy and other data needed to formulate soil and crop management alternatives and determine technical advice and credit needs.

Fluctuations in production as a result of insect pest population instability, unreliable rainfall or plant disease are more critical in the case of small farms than large ones, from the standpoint of the ability of the enterprise to absorb setbacks. Where such fluctuations can be anticipated special arrangements are called for—such as crop insurance—in order to aid the smaller units in times of crop loss. An example of such a policy is insurance against hurricane damage extended to Puerto Rican coffee producers.

* See McLoughlin (1970) for model studies of food production in Africa. For agronomic research results, also pertinent to subsistence farming in tropical Africa, see Jurion, F., and Henry, J. (1969). The authors are former directors of the 'Institut national pour l'étude agronomique du Congo' (INEAC).

(4) REFERENCES

Allan, William (1965). *The African Husbandman.* Barnes-Noble, New York, 505 pp.

Allee, W. C., and Schmidt, Karl P. (1951). *Ecological Animal Geography.* John Wiley and Sons, Inc., New York. Second edition. 715 pp.

Belshaw, Michael H. (1967). *A Village Economy: Land and People of Huecorio.* Columbia University Press, N.Y. 421 pp.

Bigger, M. (1969). Giant looper, *Ascotis selenaria reciprocaria* Walk., in Tanzania. *East African Agricultural and Forestry J.,* **35**, No. 1.

Billings, Martin H., and Arjan, Singh (1971). The effect of technology on farm employment in India. *Development Digest,* **9**, No. 1.

Boza-Barducci, Teodoro (1972). Ecological consequences of pesticides used for the control of cotton insects in Canete Valley, Peru. In *The Careless Technology: Ecology and International Development.* John P. Milton and M. T. Farvar, eds. Doubleday and Co., Natural History Press, New York.

Bradfield, Richard (1970). Increasing food production in the tropics by multiple cropping. In *Research for the World Food Crisis.* D. G. Aldrich, ed. Amer. Assoc. for the Adv. of Sci., Washington, D.C. (Publication No. 92).

Brookfield, W. D. (1961). The highland peoples of New Guinea: a study of distribution and localization. *Geographical J.,* **127**.

Brown, A. W. A. (1957). The spread of insecticide resistance in pest species. *Advances in Pest Control Research,* Vol. II, R. L. Metcalf (ed.), Interscience, New York.

Brown, L. R. (1970). *Seeds of Change, the Green Revolution and Development in the 1970s.* Praeger, New York and London. 205 pp.

Carneiro, Robert L. (1956). Slash and burn agriculture: a closer look at its implications for settlement patterns. In *Men and Cultures* (Selected papers of the Fifth International Congress of Anthropological and Ethnological Sciences). Anthony F. C. Wallace, ed.

Carneiro, Robert L. (1961). Slash and burn cultivation among the Kiukuru and its implications for cultural development in the Amazon Basin. In *Antropologica Supplement No. 2* (Proceedings of a symposium on: The Evolution of Horticultural Systems in Native South America). Caracas. Anthony F. C. Wallace, ed.

CCTA/CSA (1960). *Colloque CCTA/FAO sur le Quélea.* Publication No. 58: Commission de coopération technique en Afrique au sud du Sahara/Conseil scientifique pour l'Afrique au sud du Sahara. Proceedings of a symposium held at Bamako, Sudan Republic, May 1960. 190 pp.

Chisholm, Michael (1967). *Rural Settlement and Land Use: An essay in Location.* John Wiley and Sons, New York. (Science editions, paperback). 207 pp.

CIDA, Comite Interamericano de Desarrollo Agricola (1965). *Tenencia de la Tierra y Desarrollo Socio-economico del Sector Agricola: Guatemala.* (1966). *idem.: Peru.* Pan American Union, Washington, D.C. 244 pp. and 496 pp.

Com-Haire, M. (1964). Blue-green algae in rice fields. *Agri-Digest,* No. 3.

Coons, G. H. (1953). Breeding resistance in crop plants. In *Plant Diseases (1953 Yearbook of Agriculture).* U.S. Department of Agriculture, Government Printing Office, Washington, D.C.

Cooper, Charles F. (1969). Ecosystem models in watershed management. In *The Ecosystem Concept in Natural Resource Management.* George M. Van Dyne, ed., Academic Press, New York, 383 pp.

Cummings, Ralph W. (1968). Wheat production prospects in India. *Development Digest,* **6**, No. 3.

Davies, J. E., *et al.* (1968). Pesticides in people. *Pesticides Monitoring J.,* **2** (**2**), 80–5.

De Coene, R. (1956). Agricultural settlement schemes in the Belgian Congo. *Tropical Agriculture,* **33** (**1**), 1–12.

Delwiche, C. C. (1970). Nitrogen and future food requirements. In *Research for the World Food Crisis.* Daniel G. Aldrich, ed., American Association for the Advancement of Science, Washington, D.C. (Publication No. 92).

Donald, L. (1970). Food production by the Yalunka household, Sierra Leone. In *African Food Production Systems.* P. F. M. McLoughlin, ed., Johns Hopkins, Baltimore, Maryland.

Dyson-Hudson, N. and R. (1970). The food production system of a semi-nomadic society: the Karimojong, Uganda. In *African Food Production Systems.* P. F. M. McLoughlin, ed. Johns Hopkins, Baltimore, Maryland.

Edmundson, W. F., *et al.* (1968). Dieldrin storage levels in necropsy adipose tissue from a South Florida population. *Pesticides Monitoring J.,* **2** (**2**), 86–9.

Evans, Francis C. (1956). Ecosystem as the basic unity of ecology. *Science,* **123**, 1127–8.

FAO Fisheries Division. (1957). The Tilapias and their culture: a second review and bibliography. *FAO Fisheries Bulletin,* Vol. 10, No. 1.

Floyd, Barry and Adinde, Monica (1967). Farm settlements in Eastern Nigeria: a geographical appraisal. *Economic Geography,* **43 (3)**, 189–230.

Frank, R., *et al.* (1970). Chlorinated hydrocarbon residues in the milk supply of Ontario, Canada. *Pesticides Monitoring J.,* **4**, No. 1.

Freeman, Peter H. (1963). *Some Factors affecting Land Use in Chinchero, Peru.* Inter-American Institute of Agricultural Sciences, Turrialba, Costa Rica, 132 pp.

Gillet, J. W., ed. (1970). *The Biological Impact of Pesticides in the Environment.* Oregon State University Press, Corvallis.

Haissman, I. (1971). Generating skilled manpower for irrigation projects in developing countries: a study of North-West Mexico. *Water Resources Research,* **7** (**1**), 1–17.

Harris, C. R., and Sans, W. W. (1969). Absorption of organochlorine insecticide residues from agricultural soils by crops used for animal feed. *Pesticides Monitoring J.,* **3** (**3**), 182–5.

Harrison, H. L., *et al.* (1970). Systems studies of DDT transport. *Science,* **170**, No. 503.

Haviland, G. L. (1969). Rice in India: promising varieties still face problems. *Foreign Agriculture,* December.

Hoyle, J. C. (1969). The effect of herbicides on the growth of young coconut plants. *Tropical Agriculture,* **46** (**2**), 137–43.

Hunter, John M. (1967). Population pressure in a part of the West African savanna: a study of Nangodi, North-East Ghana. *Ann. Assoc. Amer. Geog.,* **57** (**1**), 101–149.

Hunter, John M. (1967). The social roots of dispersed settlement in Northern Ghana. *Ann. Assoc. Amer. Geog.,* **57** (**2**), 338–49.

Johnston, Bruce F., and Mellor, J. W. (1961). The role of agriculture in economic development. *The American Economic Review*, **51** (**4**), 566–93.
Jurion, F., and Henry J. (1969). *Can primitive farming be modernized?* Wellens-Pay S.A., Bruxelles. 457 pp. (translated from *De l'agriculture itinerante à l'agriculture intensifiée*, published by INEAC, 1967).
Kanel, Don (1967). Size of farm and economic development. *Indian J. Agricultural Economics*, **22** (**2**), 26–44.
Keefer, J. F. (1969). An afterlook at the Philippine rice breakthrough. *Foreign Agriculture*, March. pp. 4–5.
Kellog, C. E. (1949). *An Exploratory Study of Soil Groups in the Belgian Congo.* The National Institute for the Study of Agronomy in the Belgian Congo (Scientific Series No. 46), pp. 61–71.
Kok, L. T. (1972). Toxicity to tropical fish in the rice paddies by insecticides used for Asiatic rice borer control. In *The Careless Technology: Ecology and International Development.* John P. Milton and M. T. Farvar, eds. Doubleday and Co., Natural History Press, New York.
Linares de Sapir, O. (1970). Agriculture and Diola society, In *African Food Production Systems.* P. F. M. McLoughlin, ed., John Hopkins, Baltimore, Maryland.
Macek, K. J. (1970). Biological magnification of pesticide residues in food chains. In *The Biological Impact of Pesticides in the Environment.* James W. Gillet, ed., Oregon State University Press, Corvallis, Oregon, 210 pp.
McLoughlin, P. F. M., ed. (1970). *African Food Production Systems.* Johns Hopkins, Baltimore, Maryland.
McNew, G. L. (1962). Pest control in relation to human society. Paper published in the *Proceedings of a Symposium entitled New Developments and Problems in the Use of Pesticides.* National Academy of Sciences—National Research Council, Washington, D.C. (Publication No. 1082).
McNew, G. L. (1966). Progress in the battle against plant disease. In *Scientific Aspects of Pest Control.* National Academy of Sciences, Washington, D.C., pp. 73–101.
May, Jacques M. (1961). *Studies in Disease Ecology*, Vol. 2. Hafner, New York.
Metcalf, R. L. (1958). *Advances in Pest Control Research.* Interscience, New York. Two volumes.
Moseman, A. H. (1966). Pest control—its role in the United States economy and in the world. In *Scientific Aspects of Pest Control.* National Academy of Sciences, Washington, D.C., pp. 26–38.
Moubry, R. J., *et al.* (1968). Dieldrin residue in an orchard-dairy area of Wisconsin. *Pesticides Monitoring J.*, **2**, No. 1.
Myren, Delbert T. (1970). The Rockefeller Foundation Programme in corn and wheat in Mexico. In *Subsistence Agriculture and Economic Development.* C. R. Wharton, ed. Aldine, Chicago.
National Academy of Sciences (1971). *Chlorinated Hydrocarbons in the Ocean Environment.* NAS, Washington, D.C.
Negherbon, William O. (1959). *Handbook of Toxicology, Vol. III.* Insecticides, a compendium. Dayton, Ohio. (Wright Air Development Centre Technical Report, 55–16).
Newsom, Dale (1972). Synthetic organic insecticides for control of agricultural pests in Louisiana. In *The Careless Technology: Ecology and International Development.* Milton and Farvar, eds., Doubleday & Co., New York.

Nye, P. H., and Greenland, D. J. (1960). *The Soil under Shifting Cultivation.* Commonwealth Agricultural Bureaux, Farnham Royal, Bucks., England. 156 pp. (Technical Communication No. 51).

Ochse, J. J., *et al.* (1961). *Tropical and Subtropical Agriculture.* Macmillan, London. Two volumes.

Official Report of the United States Delegation to the FAO International Conference on Weed Control (1970).

Oltman, R. E., *et al.* (1963). *Amazon River Investigations Reconnaissance Measurements of July, 1963.* U.S. Geological Survey, Washington, D.C. (Geological Survey Circular No. 486).

Owen, D. F. (1966). *Animal Ecology in Tropical Africa.* W. F. Freeman, San Francisco.

Panel on Herbicides of the President's Science Advisory Committee (1971). *Report on 2, 4, 5–T.* U.S. Govt. Printing Office, Washington, D.C.

Plath, C. V. and van der Sluis, Arjen (1968). *Use Potencial de la Tierra: Parte VII Istmo Centroamericano.* FAO, Rome, 30 pp.

Poleman, Thomas T. (1964). *The Papaloapan Project: Agricultural Development in the Mexican Tropics.* Stanford University Press, Stanford.

Prescott, J. R. V. (1961). Overpopulation and overstocking in the areas of Matabeleland. *Geographical J.*, **127**.

Rees, A. M. Morgan (1966). The economics of tropical grassland. *Tropical Pastures.* William Davies and C. L. Skidmore, eds., Faber and Faber, London.

Reining, C. (1970). Zande subsistence and food production. In *African Food Production Systems.* P. F. M. McLoughlin, ed. Johns Hopkins, Baltimore, Maryland.

Reining, Priscilla (1970). Social factors and food production in an East African peasant society: the Haya. In *African Food Production Systems.* P. F. M. McLoughlin, ed. Johns Hopkins, Baltimore, Maryland.

Reining, Priscilla (1965). Land resources of the Haya. In *Ecology and Economic Development in Tropical Africa.* D. Brokenshaw, ed., Institute of International Studies, Berkeley, California. (Research Series No. 9).

Revzin, A. M. (1970). Some acute and chronic effects of endrin on the brains of pigeons and monkeys. In *The Biological Impact of Pesticides in the Environment.* James W. Gillet, ed. Oregon State University Press, Corvallis.

Reynolds, H. T. (1958). Research advances in seed and soil treatment with systemic and nonsystemic insecticides. In *Advances in Pest Control Research, Vol. 2.* R. L. Metcalf, ed. Interscience, London and New York.

Richards, P. W. (1964). *The Tropical Rainforest: An Ecological Study.* University Press, Cambridge, England. 450 pp.

Ridker, Ronald G. (1971). Agricultural mechanization in South Asia. *Development Digest,* **9**, No. 1.

Rolls, E. C. (1969). *They All ran Wild: The Story of Pests in Australia.* Angus and Robertson, Sydney.

Round Table (1969). Genetic dangers in the green revolution. *Ceres,* **2**, No. 5.

Rutherford, John (1966). Double cropping of wet padi in Penang, Malaya. *The Geographical Review,* April. pp. 239–255.

Shaw, Robert D'A. (1971). The employment implications of the green revolution. *Development Digest,* **9**, No. 1.

Smallman, B. N. (1965). Integrated pest control. *Australian J. Science,* **38**, 230–44.

Smith, R. F., and van den Bosch, R. (1967). Integrated control. *Pest Control: Biological, Physical and Selected Chemical Methods.* W. W. Kilgore and R. L. Doutt, eds., Academic Press, New York.

Southwood, T. R. E., and Way, M. J. (1970). Ecological background to pest management. In *Concepts in Pest Management.* R. L. Rabb and F. E. Guthrie, eds., North Carolina State University, Raleigh.

Stakman, E. C. (1967). *Campaigns against Hunger.* Belknap Press, Cambridge, Massachusetts.

Stickel, Lucille F. (1968). *Organochlorine Pesticides in the Environment.* U.S. Department of the Interior, Fish and Wildlife Service Special Scientific Report—Wildlife No. 119. Government Printing Office, Washington, D.C.

Tatum, L. A. (1971). The southern corn leaf blight epidemic. *Science*, **171** **(3976)**, 1113–6.

U.S. Department of the Interior, Fish and Wildlife Service (1965). *The Effects of Pesticides on Fish and Wildlife.* Fish and Wildlife Circular, No. 226. Government Printing Office, Washington, D.C.

USAID Mission to the Philippines 1968. Do-it-yourself rice kits. *Development Digest,* **6**, No. 3.

Van der Meer, Canute (1967). Population patterns on the Island of Cebu, Philippines. *Ann. Assoc. Amer. Geog.,* **57**, (**2**) 315–38.

Van der Plank, J. E. (1963). *Plant Diseases: Epidemics and Control.* Academic Press, New York.

Van Dyne, George M., ed. (1969). *The Ecosystem Concept in Natural Resource Management.* Academic Press, New York, 383 pp.

Van Middelem, C. H. (1969). General summary and conclusions; residues in food and feed. *Pesticides Monitoring J.,* **3**, (**2**), 100–1.

Waggoner, P. E., and Horsfall, J. G. (1969). *EPIDEM; A Simulator of Plant Disease written for a Computer.* (Bulletin No. 698 Connecticut Agricultural Experimental Station), New Haven, Connecticut, April 1969.

Wagley, Charles (1953). *Amazon Town: A Study of Man in the Tropics.* MacMillan Co., New York.

Walters, Harry E. (1970). Land, labour and the new seeds in South-East Asia. *Foreign Agriculture,* **8**, No. 46.

Westing, A. H. (1971). Ecocide in Indochina. *Natural History,* **130**, (**3**), 56–61.

Wharton, C. R., ed. (1970). *Subsistence Agriculture and Economic Development.* Aldine, Chicago.

Whiteside, Thomas (1969). *Defoliation.* Ballantine, New York.

Wittwer, S. H. (1970). Research and technology on the United States food supply. In *Research for the World Food Crisis.* Daniel G. Aldrich, ed. Amer. Assoc. Adv. of Science, Washington, D.C., (Publ. No. 92).

Wright, A. C. S. (1960). Land classification in the islands of the South-West Pacific. *Proceedings of the Ninth Pacific Science Congress,* Vol. 18 (Soil and Land Classification). Bangkok. pp. 44–9.

Wurster, Charles F. (1968). DDT reduces photosynthesis by marine phytoplankton. *Science*, **159**, No. 1474.

CHAPTER 7

River Basin Development Projects

(1) INTRODUCTION

As the pace of development accelerates in all the continents of the world, the pressure to make optimum use of available freshwater resources has intensified. Technologies and management techniques for river basin development have undergone rapid growth to meet the demands of rising populations. The range of water-based requirements is broad, including agriculture, industrial and domestic needs, energy, fisheries, transportation, flood control and recreation. Increasingly, plans for integrated development, often involving multinational cooperation, have been elaborated with the aim of optimum development of entire river basins. The latter tend to form relatively cohesive hydrological and ecological systems, although political, economic and cultural boundaries may enclose only small portions of an entire basin or, alternatively, much larger systems of influence than a single river drainage.

The majority of river development programmes have tended to concentrate on reservoir and irrigation projects. For this reason, this chapter will focus on these two forms of development, and attempt to suggest ecological principles based on past experience, that should be applicable to future planning and management. An examination of the ecological issues related to hydropower and other energy sources should certainly be made, but falls outside the scope of the chapter.

Gilbert White (1972) recently summarized the urgent need for an ecological reorientation of river basin projects (Figure 26):

'A puzzling aspect of many development projects is why they are not accompanied by more searching scientific investigation of their ecological consequences. To what conditions can we trace the lack of attention given to fisheries studies in a hydroelectric reservoir project? The same question can be directed to irrigation projects and schistosomiasis, flood control schemes and soils, the effects of pesticides or fire control, and a host of other relationships... Whatever the corrective measures, we are not doing conspicuously well with them. There is good reason to think that development projects are spreading faster than efforts to anticipate their full consequences.'

As in other kinds of development which employ new technologies or produce major environmental alterations, the ecological effects of river basin development projects range from the immediate, very specific and readily predictable impacts to the less specific, less easily discerned, long term and possibly unforeseeable impacts. Cause and effect linkages parallel the spectrum of predictable to unpredictable consequences of controlling rivers and diverting water. Perhaps the most difficult task facing the planner is

Figure 26. 'A puzzling aspect of many development projects is why they are not accompanied by more searching scientific investigation of their ecological consequences.' Kariba Dam, Zambesi River. (Photograph by S. T. Darks: courtesy Ministry of Information, Rhodesia).

that of assigning ecological, social and economic costs and benefits to environmental changes, associated with river basin development but lacking any obvious causal relationship to the actual construction of a dam or of an irrigation system. In the following sections dealing with the various types of ecological impact associated with river basin development, potential and relatively predictable impacts on aquatic life, including fish, aquatic plants and organisms which transmit human disease, are treated first. The ecological role of watershed vegetation is presented in terms of soil erosion and runoff control. The scientific and recreational values of resources that may be lost in the process of river basin development are explored, and ecological problems associated with resettlement are discussed. Special ecological problems inherent in geological changes and irrigation in arid environments, also receive attention.

The individual problems discussed have been the objects of considerable investigation by scientists, and what is presented is not new material. However, it is hoped that this collective treatment of the various direct and indirect ecological consequences of river basin development will provide the planner with a framework for more adequately assessing the total costs and benefits of a project and for better anticipating the types of information and management needed to minimize adverse environmental impacts and their associated socioeconomic costs. The emphasis throughout is on tropical river ecosystems, which involve specially complex problems.

(2) PHYSICAL ASPECTS

Over three-quarters of the world's land surface has been estimated as potentially available to river basin development (United Nations, 1970). Those terrestrial areas not susceptible are regions either too dry or too cold. Greenland, Antarctica, and northernmost North America and Eurasia are among the places considered too cold; in the deserts of Africa, central Asia, Australia, the Middle East and western South America high evapotranspiration rates, salinization and small, inconsistent stream flows limit river basin projects.

In some cases, however, water resources development in these regions is possible. Hydropower and navigation projects are occasionally feasible in cold environments, even though agriculture may be limited. Likewise, arid regions may be traversed by streams originating elsewhere, making integrated river basin projects potentially possible. In other arid regions, groundwater reservoirs may enable grazing, agricultural and other development activities to be undertaken.

The biosphere's water cycle involves a constant interchange between the world's water bodies, as shown in Figure 27.

Rivers (and lakes) play an important role in the hydrologic system of the biosphere. Together with groundwater supplies, they drain off to the oceans the land surfaces' excess precipitated freshwater not evaporated back into the atmosphere. Although this part of the water cycle is only a small fraction of the total biospheric water supply, it is an extremely vital one to terrestrial and freshwater ecosystems, and to man.

Rivers and streams are particularly important agents of land erosion, transport and deposition. Together with biological, climatic and geological factors, rivers have helped to determine soil characteristics and influenced water supplies available to ecosystems. Biological systems have, through evolution, adapted to the wide range of ecological conditions in river basins, and man's cultures have also been intimately linked to the constraints and opportunities within river systems. Many of the most successful agriculturally based human societies have benefited from the abundant seasonal water supplies and rich sediments provided by rivers. At the same time that man has benefited from utilizing the soils conditioned by water supplied by river basins, human culture has often been constrained by their processes of basin erosion, salinization and flooding.

Ever since the origin of agriculture, communities have attempted to control and manage the supply, distribution and quality of available water. Irrigation networks, barrages, pond construction and soil terracing have long been practised in an attempt to increase the productive capacities of river basins. Many of these early attempts were successful. In other cases,

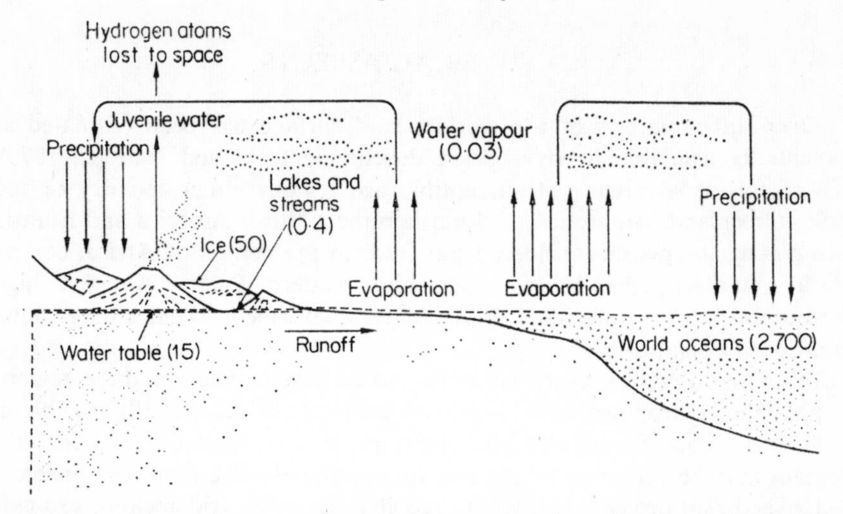

Figure 27. Water cycle in the biosphere requires that worldwide evaporation and precipitation be equal; hydrogen losses to space are presumably replaced by juvenile water. Ocean evaporation, however, is greater than return precipitation: the reverse is true of the land. Excess land precipitation may end up in ice caps and glaciers that contain 75 per cent of all fresh water, may replenish supplies taken from the water table by transpiring plants or may enter lakes and rivers, eventually returning to the sea as runoff. Numbers show minimum estimates of the amount of water present in each reserve, expressed as a depth in metres per unit area of the earth's surface. (From H. L. Penman, 'The water cycle'. Copyright © 1970 by *Scientific American* Inc. All rights reserved).

particularly in arid lands, river control techniques resulted in salinization, waterlogging, disruption of fisheries, heightened erosion and increased waterborne disease.

Recently, a number of new techniques have made it possible to consider river basin development projects on a scale formerly impossible. Approximately 10 per cent of the world's total annual stream flow is now regulated by man. It is estimated that by the year 2000 this regulated flow will have increased to two-thirds of the total stream flow (Szestay, 1972). At the same time, demands of growing populations for food, electric power, and economic growth have led to intensive pressure to initiate rapid, large-scale river development. Many of the new methods of achieving this (such as large dams) were first conceived and applied in highly developed countries of the temperate zone. The application of these engineering works to arid and tropical environments, however, has often led to serious adverse social and ecological consequences that were not anticipated by planners (Farvar and Milton, 1972).

(3) PREVENTIVE PLANNING

Planning, and the related studies upon which it is based, is the first and perhaps most important step in avoiding unwanted consequences of a river basin project, whether it be a dam or an irrigation scheme. Preventive planning is basically the incorporation into a plan of all foreseeable consequences, environmental, economic and social, of a development activity, not just the immediate consequences as defined by the specific goal of the activity. Thus the immediate goal of a dam may be to elevate water and harness its kinetic energy to generate electric power. In addition to the immediate consequence—that of creating a reservoir and generating electricity—the planner should anticipate the upstream and downstream effects of an artificial lake and of regulating water flow. The possible effects are numerous and are discussed later in this chapter.

Failure to anticipate and evaluate adequately the various impacts of such projects has resulted in an incomplete assessment of costs and benefits. Corrective measures—which represent real costs—are now being taken in several tropical dam projects to solve problems of human disease, fisheries, erosion, sedimentation and aquatic weeds. In some instances, the value of resources lost has become apparent only after the project has been completed.

Comprehensive resource surveys and evaluations are therefore required in order to prevent mistakes so often made in the past. Multi-purpose, integrated planning, encompassing entire river basins, is coming to be seen as a prime prerequisite for successful individual river basin projects. The integrated resource surveys of the Plata River Basin undertaken by the OAS exemplifies this planning approach (OAS, 1969). The Preparatory Commission for the Volta Project also undertook excellent comprehensive surveys aimed at forecasting the enviromental impacts. Concurrently, numerous recent research efforts have aimed at elucidating the basic ecological factors necessary to successful integrated basin planning. This chapter attempts to summarize some of the most important of these and to suggest a series of concepts that should be helpful to future planning. In the application of these principles, however, several summary points deserve emphasis (in relative order of project timing):

1. Comprehensive, long-term evaluations of the social and ecological impacts of past river basin development projects ought to be initiated by national and international agencies responsible for development. Such evaluation should include global inventory and integration of information on existing man-made lakes and irrigation projects.
2. Long-term environmental research and monitoring, both basic and

applied, should be undertaken in river basins likely to undergo development. Such information is often critical in the identification of successful development techniques and alternatives.

3. Local environmental monitoring of natural and modified systems ought to carry through all phases of pre- and post-project development. Particular attention to possible environmental disruptions should be emphasized in this monitoring.

4. Ecological surveys and studies to anticipate the impact on the aquatic and terrestrial environment ought to be initiated at the earliest possible phase of river basin development planning.

5. Wherever possible, specific water control ought to be part of a comprehensive and integrated river basin plan that includes ecological considerations. This integrated plan, in turn, should be harmonized with overall national political and economic goals.

6. Specific ecological guidelines should be formulated on the basis of pre-project research and on-going monitoring for pre-investment survey criteria; cost-benefit project analysis criteria; project construction design; and project management and evaluation studies.

7. Timing of project planning, construction and follow-up should be adequate to allow full incorporation of ecological considerations, data and findings.

8. Project studies, from early phases of identification to later stages of assessing feasibility and planning, should include investigation of local social and cultural systems in relation to the national society and culture. The aim would be to compare current patterns of behaviour and values with patterns which seem more congenial to an economically developed river basin. Thus a project would be planned to suit the current and potential adaptive capabilities of people who have their own way of life rather than some elaborate, foreign form of technology whose introduction might eventually stimulate more costs than benefits. In view of the complex changes often involved in river basin projects, it is just as essential to study and plan their adaptation to local cultural patterns as it is to weigh the local ecological, economic and engineering factors.

9. Where local populations must be adversely affected by development, alternative development plans should be initiated or adequate arrangements made for full compensation for losses, retraining, environmentally sound relocation and other needs.

10. If costly environmental impacts are anticipated, alternative plans should be formulated in conjunction with the primary development plan and their environmental impacts, in turn, should be taken into account.

11. Estimates of all the ecological costs and benefits of river basin development projects (including alternatives) should be included in overall cost-benefit analyses.

12. Adequate provision for ecological aspects should be made in the final project funding, training, management and extension services. Where necessary, aid should be given to local agencies responsible for development but lacking ecological competence, to enable them to enlist trained ecologists to assist in planning and management.

Areas of expertise required for the study of possible ecological impacts of river basin development include: ecosystem ecology, limnology, parasitology, nutrition, sanitation, ichthyology, entomology, ornithology, botany, marine and freshwater biology, geomorphology, and hydrology. The ecological interpretation of the data supplied by these disciplines would cover:

1. energy pathways and nutrient cycling in river basin environments prior to development;
2. ecology of plant communities in terrestrial and aquatic environments prior to development;
3. ecology of major human communities related to the development;
4. ecology of reservoir ecosystems;
5. ecology of irrigation ecosystems; and
6. identification of ecological factors relevant to river basin planning, particularly emphasizing fisheries, public health, aquatic weeds, forestry, salinization, waterlogging, water quality, recreation, tourism and protected areas.

(4) MAN-MADE LAKES AND RIVER BASIN ECOLOGY

(a) Benefits from Reservoirs

One of the most common types of river basin development is the construction of man-made lakes. Although the engineering of a project may be relatively straightforward, the interrelationships of the system involved are almost always complex (Figure 28).

Reservoirs usually are built for some primary purpose, such as hydropower, irrigation or flood control. In many cases, an integrated series of benefits are projected that include all three of these purposes as well as other benefits such as reservoir fisheries, transportation improvements, provision of domestic and industrial water supplies, and recreational facilities. Although we are mainly concerned here with the more important ecological problems arising from primary development goals, it needs to be emphasized that most development projects are themselves intended to *solve* significant problems. The difficulties that arise are usually due to lack of attention to

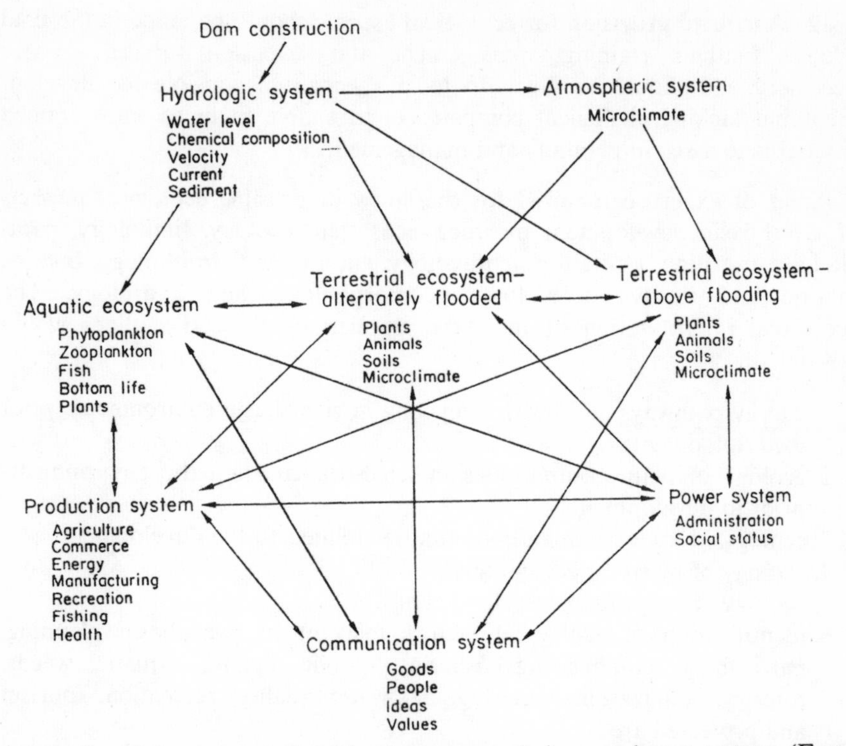

Figure 28. Major interrelationships in a reservoir impoundment system. (From White, 1972).

secondary problems and costs that have often undermined the effectiveness of the original development objectives.

The provision of hydropower can be of great significance to developing nations. Often there are few other practical alternatives to the growing demands for an abundant, inexpensive energy source. Hydropower helps diminish the need to purchase fossil or atomic fuels and, thereby, reduces import requirements of countries which are short of the necessary funds. Similarly, use of hydropower helps reduce environmental costs by preventing the build-up of air and water pollution problems linked to fossil fuels and to radioactive waste disposal and thermal discharges from atomic plants.

Multi-purpose reservoirs can offer a wide range of other major productive benefits and tend towards more favourable cost/benefit analysis than single purpose ones. Thus water stored in man-made lakes can provide supplementary water for extending growing seasons in already irrigated areas or can make it possible to irrigate new lands. Dams can also help increase agricultural production by controlling destructive floods and, through stabilizing downstream water levels, facilitate transportation to markets. In addition, well-managed man-made lakes can provide valuable new fisheries and, in some cases, an expanded economic potential for lake-based tourism. Finally, the storage and stabilization of river basin water can help increase both domestic and industrial water supply. Taken together, these benefits often combine persuasively to encourage construction of man-made lakes. The primary benefits are usually carefully evaluated and quantified in the original costing of a project. On the other hand, secondary problems and costs are often inadequately evaluated, leaving serious gaps in the project appraisal. The result is that the case for constructing a reservoir is often stressed at the expense of possible alternative development patterns that take better account of the ecological and social factors. In view of this tendency, the various secondary impacts will now be discussed in detail.

(b) Fisheries Development Issues

Reservoir fisheries are normally established for commercial subsistence or recreational purposes, or for a combination of them. Under careful management, the three kinds of fishing can complement each other and expand the total benefits obtained.

To achieve maximum production, detailed evaluation of fishery problems and prospects must be initiated early in the planning process. Studies of the natural hydrobiology, existing and potential fish utilization, effects of engineering works and of various river basin land and water uses, and techniques to offset potential fishery losses (such as sluice gates, fish passes, and well-regulated water regimes), ought to be included in such evaluations. In addition, the total gains from potential reservoir production should be weighed against potential production losses in the free-flowing river system.

Particularly where fish production is a primary goal of a reservoir, other fishery development strategies ought to be investigated as alternatives to the creation of man-made lake fisheries. Lastly, pre-impoundment studies for projects in tropical regions would certainly be strengthened by encouraging long-term research and monitoring of tropical fish fauna, Much of the information on this fauna, which is vitally needed for effective management, is not yet known.

(i) Fish Migration and Downstream Impacts

One of the commonest problems arising in the construction of reservoirs is that dams act as barriers to the movement of fish up and downstream. Many fish have migratory life cycles that require movement to different parts of the river system, usually to reach feeding or spawning grounds. When a dam blocks the completion of this cycle, the fishery can be eliminated or severely reduced. Sometimes fish attempting to pass downstream may be destroyed in turbines installed below a dam.

These adverse consequences can sometimes be mitigated through careful study of the ecology and behaviour of fish stocks, followed by construction of fish lifts, fish locks, fish passes, fish traps, and guidance devices to pass fish in both directions over dams. In other cases, study of the total situation has shown that fish passes are not needed.

An additional problem is that dams usually alter the natural water regime downstream. Fish often depend upon seasonal shifts in water flow to migrate into areas beyond the normal river edge for feeding and spawning. Reservoir construction can prevent high enough flows, or cause poorly-timed flows that block these movements. Sometimes flood regulation that only prevents extremely high flows can forestall spawning mortality, but in most cases the impact on lateral fish movement has adverse consequences.

Other downstream changes which may adversely affect fish populations include alterations in water nutrient levels, temperature, dissolved oxygen levels, sedimentation, chemistry and velocity. Since much sediment and suspended organic material will normally be deposited in the reservoir, an important source of food for plankton and bottom feeders may be cut off or greatly reduced. Dams may also alter the amount of living space available to fish. Any reduction in downstream discharge can disrupt feeding, spawning and nursery grounds, and cause greater exposure to predation. With a reduced discharge, the stream's capacity for self-purification (particularly through oxidation) can also be degraded, leading again to fishery losses. Large sudden alterations in flow can be equally or more damaging.

Another downstream problem often caused by man-made lakes is the shifts in salinity and turbidity, or chemical changes in estuarine and adjacent marine areas. These alterations can have important impacts on marine fish and shellfish. Thus the construction of the Aswan Dam is reported to have brought about the collapse of a major sardine fishery which formerly accounted for upwards of half Egypt's total marine catch; the silt which once provided nutrients for coastal plankton and other organisms on which the fish fed, is now deposited in Lake Nasser. On the other hand, this loss must be weighed against the benefits which should accrue from the creation of a major new lake fishery in Lake Nasser itself.

(*ii*) *Reservoir Fisheries*

As indicated above, adverse consequences to marine, estuarine and riverine fisheries can sometimes be offset by the development of successful reservoir fisheries. Nevertheless, there are often serious problems to be met in establishing productive fisheries in man-made lakes. Most of them start from a natural stock of fishes ecologically adapted to riverine conditions. Requirements of such species for spawning, food, water temperature, current and oxygen are likely to be poorly met in reservoirs. However, riverine species that prefer quiet water, for example, will probably do better than those needing current. Similarly, although few riverine fishes are plankton feeding, those that are will be favoured in a lake environment.

Because of these adaptation problems, new reservoirs that have not been carefully studied with a view to management and stocking with species suited to lacustrine situations, will tend to be under-utilized by native riverine fish species (Jackson, 1966).

There may of course be numerous new niches created in the reservoir which none of the native species can occupy. On the other hand, introduction of exotic species may only be of limited success and, in some cases, can cause unanticipated problems of infestation. Such introductions should therefore only be attempted (1) after detailed studies of the existing native fish fauna have been completed and indicate that no alternative is feasible, and (2) after intensive investigations of possible secondary ecological impacts.

Another problem for fisheries is that the flooding of a man-made lake may silt up or inundate feeding and spawning grounds. Associated marsh drainage and stream canalization can have equally adverse effects. Shifts in depth, vegetation, water velocity, and the character of the bottom are all potentially destructive fâctors. However, as in the case of other impacts, such adverse consequences may be reduced by thorough ecological studies of proposed physical alterations of the water regime at an early stage in the planning.

In most tropical reservoirs where there is relatively little seasonal fluctuation in temperature, humidity and rainfall, water temperature zones tend to become stratified and stabilized. In addition, the saturated oxygen available to nurture aquatic life is liable to be limited to surface waters. Deeper zones are normally deficient in oxygen, which limits fish life and makes these zones unproductive. The problems can be magnified where hydrogen sulphide gas develops from decomposing vegetation at deeper levels and can be quite destructive to fish life (Lagler, 1969).

By contrast, most rivers tend to have relatively uniform temperature and dissolved oxygen levels from surface to bottom, conditions which are usually more favourable to various aquatic species.

Problems of reservoir stratification can sometimes be mitigated through careful location of water outlets. Other possibilities are diffused air systems, multi-level intakes and withdrawal structures; however, these techniques seem best adapted to small impoundments and of limited use in large, deep reservoirs. Again, intensive prior study of the aquatic ecosystem is a necessary requisite to the success of remedial measures.

During the first few years following inundation, reservoir fisheries will benefit from the initial rapid release of nutrients from decomposition of inundated vegetation and soils. This high fertility is reflected in a temporary rise in initial fish production. Normally a decline in production follows as decomposition and nutrient relase slow down, and as aquatic weeds start to take up some of the available nutrients.

Reservoirs characteristically undergo this pattern of an initial increase in production followed by a steady decline, but at some later date following this 'trophic depression', production usually stabilizes at a slightly revived level, though not as high as during the initial outburst.

The time taken for a reservoir to stabilize depends very much on latitude, lakes with higher temperatures stabilizing much faster than colder ones. In colder lakes of high latitudes, where fish growth is slow, it takes many years for fish populations to build up, and care must be taken not to fish the stocks of desirable fish species until this has occurred. In tropical lakes, fish mature much faster, often within a year or so of an impoundment, and fisheries tend to be developed during the initial outburst of productivity. It is important to realize, however, when assessing capital outlay for fishing gear, that the high catches in the first few years of the fishery are bound to decline as the nutrients resulting from the flooding of new land are used up.

The increase in predacious fish populations in new reservoirs can also be a limiting factor in the production of plankton-feeders and other prey species. Because of this, it is important not to promote the expansion of lake fisheries to levels based on initial high production. Careful early study and management can help offset proliferation of undesirable predacious fish. In some cases predacious species are, however, desirable food fishes.

Other problems come from the often severe drawdown typical around reservoir shores. This can destroy spawning grounds, feeding areas, and the vegetation that supports these requirements along shorelines (Jackson, 1966). One of the more serious problems in the tropics is the spread of aquatic weeds. In some cases, these compete with fish for living space, while floating weeds cut down on available oxygen and reduce plankton production. In other situations, the aquatic plants provide shelter and increased areas for growth of flood organisms. In nearly all cases, however, waterweed invasions conflict with the ability to fish and prevent navigation. The effects of waterweed growth are discussed more fully later in the chapter. Their control by herbicides may often seem attractive, but may have dangerous side effects on oxygen level in reservoir waters, on adjacent agricultural areas, on food organisms consumed by fishes, and perhaps even on men and animals consuming contaminated fish or water. Research into specific biological controls through insects, diseases, other animals, or habitat manipulation may result in less hazardous ways to suppress weeds.

Other hazardous substances may also affect water quality and fish populations (Figure 29), such as pesticides, chemical fertilizers, sewage and industrial pollutants. Agricultural pesticide application to rice paddy water has been found to contaminate or kill fish raised for human food (Kok, 1972). Similarly, the expanding use of chemical fertilizers can stimulate waterweed and algal growth in reservoirs and primary canals, and lead to

Figure 29. Much of the value of any water development will be lost where water pollution is not checked. Dead fish in a polluted Thailand waterway. (Photograph by F. Vollmar: courtesy World Wildlife Fund).

eutrophication* of fresh waters. Pollution by sewage, silt and toxic industrial chemicals can also have severe adverse effects on fish production.

Some forms of pollution, particularly from toxic chemicals such as acids, alkalis, detergents, oils, etc., have a direct destructive effect on aquatic life. the discharge of heated water or thermal pollution can also have damaging

* The process whereby waters are enriched by nutrients (e.g. phosphates and nitrates), with consequential depletion of dissolved oxygen.

effects on aquatic systems. Other forms of pollution, such as sewage and fertilizer runoff, affect fish production by indirectly reducing the oxygen level of fresh waters.

Cases have occurred where reservoir fisheries have suffered from inadequate clearing of forest vegetation from inundation sites. The submerged vegetation often decomposes very slowly, and tree trunks and branches can be a significant hazard to fish nets and other fishing gear. Deoxygenation and toxification of water may also result. In other cases, the trees left in the inundated area may contribute to productivity by providing a substratum for fish food organisms. In any case, complete clearing of large tropical lake sites may be economically unrealistic and the technique of clearing lanes for fishing may be the best approach. Once the reservoir has been flooded, however, it may be difficult or impossible to remove submerged vegetation. Therefore, funds for study of clearing requirements (and, later on, for the actual process of removal) should be allocated early in the reservoir planning process.

Whenever a potential threat from pollution exists, good fisheries management in both rivers and reservoirs requires that adequate water quality criteria be established and enforced. Each proposed river basin development feasibility study ought to include thorough investigations of potential direct and indirect impacts on fisheries. Once the processes relevant to aquatic systems have been examined, water quality standards should be created for a variety of water uses, legislation and enforcement should support these standards, and monitoring of water quality levels should be initiated.

(*iii*) *Aquacultural Potential*

Despite the many problems encountered in maintaining river basin fisheries following the construction of man-made lakes, there is often a great potential for increasing fish production. If this potential is to be realized, it will require intensive research into the aquatic ecosystem involved, both before and after development activities have been initiated. The aim of increased production may be achieved by devising techniques to avoid losing existing reservoir and riverine production, and by developing new possibilities of obtaining higher yields from reservoirs and irrigated lands, and from methods of aquaculture, such as fish ponds. Reference has already been made to some of the ways of avoiding fishery losses; positive measures for stepping-up production are of equal importance.

One of these measures, the introduction after careful study of lake-adapted fish species to reservoirs, has already been mentioned. Another is to use a reservoir to provide water supplies for small ponds and irrigation tanks located on its periphery or downstream. Plant and animal production from these ponds may exceed both the economic and the protein yields possible from the same area under agricultural production. Depending on climatic conditions, many plants, such as water chestnut, water lettuce and watercress can be grown. A wide variety of useful animals can also be propagated, either in combination or separately. For example, many species of fish, shrimp, shellfish, other crustaceans, turtles, frogs, crocodiles, geese and ducks have been proven to be of great aquacultural value; the supplementary production of water-plants, as feed for cattle, pigs, poultry and other livestock, can be equally worthwhile.

In some cases, ponds or irrigation tanks will prove the best means to increase yields. In other situations, lands under irrigation agriculture can be utilized for simultaneous production of the irrigated crop and of various species of plants and animals. A similar aquacultural potential often exists in shallow reservoirs or along lake shorelines. Careful management of water levels may allow shore-bordering ponds to be replenished periodically with water, without interference to the filling of downstream ponds by normal gravity flow.

Whenever aquacultural projects are under study, however, great care is needed to consider related problems which might develop, such as waterweed infestation, inundation of shoreline forest and cultivation, escape of exotic species which could become pests, and the spread of snail-borne and arthropod-borne diseases.

Similarly intensive research and planning are needed to determine the social acceptability of various potential aquacultural practices, local nutritional needs and food preferences, and what provision ought to be made for education, training and management. This would include the aquacultural extension services and planning facilities needed for raising plant and animal stocks, and credit and marketing assistance to the local inhabitants. Any alternative engineering structures that may be required for aquacultural purposes must of course be carefully planned.

(c) Public Health

Reservoir construction and stabilization of irrigation systems, and the resettlement programmes, transportation improvements, and nutritional changes resulting from such river basin projects have all affected the patterns of human and domestic stock disease, especially in tropical countries.

Numerous studies have been made of this and the effects on public health. As a result, there is now ample evidence that prevention and control of disease associated with river basin development requires intensive epidemiological and nutritional investigations early in the pre-investment and feasibility surveys. This implies that specialists in disease ecology and biology, nutrition, and sanitation must be associated with development projects in good time to anticipate possible changing patterns of disease and to work out how to control them.

(*i*) *Schistosomiasis*

Perhaps the greatest current disease problem of man-made lake and irrigation projects in the tropics stems from snail-borne infections. The World Health Organization has recently identified schistosomiasis (bilharzia) as one of the largest public health threats to developing countries, probably surpassing even malaria. It seems likely that schistosomes have been human parasites for much of man's evolutionary history (Shiff, 1971). Under earlier conditions, however, the ecological relationship between parasite, the primary host (snail) and the secondary host (man) was in relative balance; human populations were usually sparse, as were the snails; relatively low levels of the disease probably prevailed.

More recently, the rapid growth of dense human populations with poor sanitary facilities, the increased mobility of infected people, and the ever more frequent construction of reservoir and perennial irrigation projects, all contributing to the excellent conditions for the fast spread of snail hosts, have made schistosomiasis the serious problem it is today. As these same factors promise to multiply the disease over the foreseeable future throughout the tropics, an integrated control approach is badly needed whenever development projects liable to aggravate schistosomiasis infection are being planned.

An excellent study of the relationship between schistosomiasis and the transition from seasonal to perennial irrigation practices has been made by Henry van der Schalie who, in a recent article (1972), writes:

'Egypt, over hundreds of years, has become unbelievably infested with bilharziasis. Health conditions in rural Egypt were documented in the delta region by John Weir and his associates in 1952 at Sindbis, near the Qalyub Egypt 10 tract. They provided first meaningful mortality and morbidity data and showed that the life expectancy of women in that region was 27 years and that of men 25. In Egypt the countryside of the delta is virtually rotten with the disease. The overpopulated area with its farm population in horribly crowded villages, the lack of sanitation and the near impossibility of building proper facilities for potable water and waste disposal, the many unfortunate daily practices that allow for an amazing exposure to infection—all contribute to make the conditions in the areas where perennial irrigation exists almost impossible to control.'

(*ii*) *Opisthorchiasis*

This is another species of snail-borne disease probably spread by expansion of perennial irrigation systems (Milton, 1969). In the Mekong Basin of Southeast Asia a liver fluke, *Opisthorchis viverrini*, is a common human parasite. Its life cycle is very similar to that of the schistosome, except that there is an intermediate stage between the snail, a species of *Bithynia*, which is host for the miracidia form, and man. The free-swimming cercaria that leave the snail, instead of directly penetrating human hosts, infect various Cyprinoid fish species. The cycle is completed when humans ingest raw or poorly cooked fish, infested with the parasite, and pass the eggs back into the aquatic system through their excreta. Inflammation and scarring of the liver, obstruction of the bile duct, cirrhosis and even cancer of the liver are linked to the disease. Undoubtedly, human resistance to many other diseases is greatly lowered by the parasitic infection.

Integrated control of these various snail-borne parasites usually combines (1) a pure drinking water supply; (2) excreta disposal methods that prevent the parasite's eggs from reaching water systems; (3) medical assistance to reduce disease incidence; (4) health education to inform people of the parasite's life cycle and of the need to follow health rules; (5) quarantine measures to halt the spread of the disease by human population movements from highly infested to less infested regions; and (6) control programmes to reduce snail populations (through molluscicides, canal water level control, and biological controls). Special care is needed in the case of molluscicide not to create severe chemical pollution problems. Snail control, through development of biological controls and management of irrigation ditch and canal levels, are usually more desirable approaches to this aspect of the problem, and need much more attention. To achieve this, careful study of natural predators on the various snail species carrying the parasite will be needed. In all cases of river basin development affecting irrigation systems, support for integrated disease control research and its subsequent application should form an essential part of the planning and execution of the project.

(*iii*) *Malaria*

Other forms of disease linked to river basin change are arthropod-borne infections: onchocerciasis, trypanosomiasis, filariasis, malaria and various viral diseases are among the most important (Lagler, 1969). Together with schistosomiasis, malaria is among the more widespread and debilitating of the public health menaces in tropical countries. Fluctuating fever, high parasite levels in the blood spleen enlargement and anaemia are common symptoms and some forms of the disease invade the brain's capillary system, causing blockage of oxygen and brain damage or death. High infection rates also often lead to considerable infant mortality, miscarriages and increased susceptibility to other diseases. Taylor and Hall (1967) indicate that infection causes significant losses in worker productivity (absenteeism), with adverse effects on the achievement of development goals. Malarial infestation of large areas of potentially productive land may hamper agriculture in regions where there is a pressing need for increased food production.

Malaria is caused by species of the genus *Plasmodium*, living in the bloodstream of the host and carried by anopheline mosquitoes: at least a hundred species of this mosquito vector are known and each has a unique ecology and life history which makes it difficult to develop single control techniques that are generally applicable. In addition, knowledge of the biology of various species is often incomplete.

As with schistosomiasis, construction of irrigation works and creation of reservoir shorelines have helped to spread malaria in many tropical areas (Hunter and Hughes, 1972). Because of the highly variable distribution and ecology of the vectors involved and the wide range of human factors influencing the disease, every river basin development programme should include health specialists with a good knowledge of local conditions. The World Health Organization, in particular, has developed numerous techniques for effective malarial control programmes and the main problems now confronting control efforts tend to be unauthorized population movements from infested areas to regions where eradication programmes are in progress, land use and settlement patterns, housing standards and water use.

Other difficulties relate to control techniques themselves. *Thus, while it is indisputable that the use of pesticides, such as DDT, particularly to keep houses free of mosquitoes, has reduced the risk of malaria in the short term, it has, in some cases, led to the development of insecticide-resistant vector strains, which increases the threat of infection. Resistance usually develops from emergence of physiological tolerance, a pattern of repellence when in contact with the pesticide, or behavioural changes which allow the vector to avoid contact with sprayed areas. In such cases, the high reproductive capacity and adaptability of the vectors through natural selection, have necessitated further expenditure on the development of new, more effective pesticides. Unless the cycle of parasite transmission from man to insect vector can be broken in other ways, a simple control approach relying on insecticides may only perpetuate the problem in a different form.*

Since river basin, reservoir and irrigation projects involve a very distinct risk of promoting the spread of malaria, particular care should be taken in planning them to study possible biological and habitat control techniques. Other less immediate and obvious control problems will later have to be considered. Thus, in cases where successful malarial control has been initiated, the reduction in infant and other human mortality may accelerate pressure on available food resources. If the rate of population increase exceeds food supplies, problems of protein and calorie deficiency diseases may simply replace the earlier problems of malarial infection.

(*iv*) *Filariasis*

Difficulties of much the same kind apply to the control of human filariasis, particularly bancroftian filariasis (*Wuchereria bancrofti*), which is commonly spread by the mosquito vector *Culex fatigans*. In areas where sewage and public health standards are low, where population densities are high and where large bodies of stagnant water prevail, the vector often quickly becomes established. Reservoir and irrigation projects that both tend to concentrate human settlement and provide good vector breeding and parasite transmission sites, therefore call for an investigation of control methods early in the planning process. The more effective methods are likely to involve a combination of mass human treatment in endemic areas and treatment of stagnant water bodies, both aimed at interrupting parasite transmission.

(*v*) *Trypanosomiasis*

Trypanosomiasis, commonly known as sleeping sickness, is an African disease caused by protozoans of the genus *Trypanosoma*. Tsetse flies, *Glossina* spp., are the insect vectors of the infection to wild and domestic animals and man. One species, *G. palpalis*, is commonly associated with forested or thicket areas bordering rivers and lakes. Large areas of Africa are still infected by the disease, which is severely debilitating and often fatal to man and his livestock.

Following construction of reservoirs in areas where the tsetse occurs, the vegetation which establishes itself along shorelines may produce a habitat favourable to the fly. Normally the total length of shoreline thus becoming conducive to its spread far exceeds that of the original tsetse habitat fringing the watercourses prior to inundation. Another factor is the tendency for human populations to be attracted by fishing opportunities in precisely that lakeside danger-zone; displaced communities may actually be resettled there; or access roads to the shore may provide another avenue for infection.

Road projects in areas where infected tsetse are present often act as a means of spreading human and bovine trypanosomes and their vectors to fly-free or uninfected zones. In such cases, movements of people and cattle, facilitated by new road networks, must be made subject to quarantine and other controls, if the spread of sleeping sickness and similarly localized diseases is to be checked.

Settlement projects have also caused problems. Colonial administrations in the past normally attempted to relocate the human population of tsetse-infested areas in fly-free hill country. But the reverse has also sometimes occurred, new settlements being sited on the more productive but infested land of riverine plains (Hunter and Hughes, 1972).

(*vi*) *Viral Infections*

Of the many arthropod-borne viral infections common in the tropics, yellow fever, haemorrhagic fever, dengue and encephalitis are among the most important. In cases where hydro-development projects are promoted, careful survey of these diseases and their potential control requirements is necessary.

Whenever water has to be stored in open containers and human populations are dense, the mosquito *Aedes aegypti* finds a favourable habitat. This species is the common vector of yellow fever, as well as haemorrhagic fever and dengue. If people whose original homes have been inundated by river basin development are resettled in upland areas with inadequate water supplies, as often happens, *Aedes aegypti*-borne infections are a likely result.

Another dangerous virus carried by mosquitoes is Japanese encephalitis. The major vector is *Culex tritaeniorhynchus,* which breeds in rice fields (particularly in Korea and Japan). In susceptible regions, where reservoirs permit extension of rice growing land and where the period of rice field inundation is lengthened, major encephalitis outbreaks may result and ought to be anticipated through control programmes (Lagler, 1969).

(*vii*) *Onchocerciasis*

More commonly referred to as river blindness, this is an infection caused by the filarial worm *Onchocerca volvulus*. The adult female worm, which can survive up to fifteen years, lives in the skin of human hosts and discharges numerous mobile microfilariae; these penetrate the skin and, if they reach the eyes, can cause permanent blindness. In some heavily infested areas, particularly in Africa, up to ten per cent of the population may be blind and whole infected valleys may be abandoned by rural peoples.

The parasitic microfilariae are transferred from man to man by the bite of a small black fly of the genus *Simulium*. The larval and pupal forms of this vector inhabit highly oxygenated, rapidly flowing streams over large areas of Africa (particularly in West and Central Africa), as well as parts of Mexico, Central America, Venezuela and Colombia.

Reservoir inundation of fast flowing streams has often had a beneficial effect by largely reducing the vector's prime habitat. For example, construction of Ghana's Volta Dam eliminated the disease from the flooded area (Waddy, 1966).

On the other hand, reservoir spillways often provide suitable breeding sites for the fly and, wherever dams are located in infested regions, submerged spillway and outlet pipes ought to be considered. Another problem is that if people from the inundated areas have to be resettled and concentrated in infested areas up or downstream of the reservoir, a higher transmission rate of the disease may be favoured.

(*viii*) *Disease Problems Associated with Population Movements*

A wide variety of communicable diseases, such as syphilis, gonorrhea, viral infections, dysentery, typhoid and cholera, are directly or indirectly stimulated by river basin development, particularly so if there is a concurrent increase of population. The construction of roads and movement of people connected with dam construction, as well as resettlement, may spread parasitic and communicable infections to new areas and new populations. Similarly, uninfected populations may be concentrated in new disease foci where there is high probability of infection. Settlers moved within reach of diseases to which they were previously unexposed may be particularly susceptible.

In development projects, particular care is needed to examine population shifts in relation to the probable alteration of their disease environment and the public health measures that will be required. Resettlement problems are particularly relevant in this connection and examined more fully in subsection (h) below.

(*ix*) *Malnutrition*

Reservoir and river basin development may cause various problems of malnutrition. In cultures highly dependent upon riverine and marine fish as a source of protein, reduction of fish stocks through interruptions of spawning and feeding, or through alterations of water nutrient levels, river volume, water quality and temperature, can be serious. Protein is already in short supply in many developing regions and further reductions may trigger serious consequences such as kwashiorkor. It is important to emphasize, on the other hand, that reservoir development may also contribute new sources of fish protein. Supplies of perennial irrigation water may also provide for increased supplies of plant protein and other important nutrients from crops. Major nutritional problems are more likely to affect resettled and some downstream communities who are unable to benefit from the new facilities.

Resettlement after reservoir construction, whether it is spontaneous or planned, often involves major changes in traditional land use patterns and diet. This is particularly true where lowland agricultural patterns on alluvial soils (such as in the wet rice regions of Asia) have to be replaced by those appropriate to upland, non-alluvial sites. Obviously, if population densities in the area surrounding the reservoir allow relocation (at suitable densities) on similar lowland sites, former land use practices and crops can usually be maintained with little change in nutritional status. But where populations are such that most of the available lowland soils are already utilized, resettlement is forced into less desirable, nutrient-poor upland areas.

When this happens, nutritional impoverishment can be caused by—(1) inability to adapt to new upland agricultural practices providing a balanced diet; (2) lack of easily available fish, on which people depended for protein before resettlement; (3) relatively poor soils that cannot sustain long-term, intensive agriculture under either subsistence or cash crops; or (4) the failure of agencies responsible for resettlement to provide for adequate credit, transportation, household and irrigation water, fertilizer, and upland crop extension services.

(d) Aquatic Weeds

Among the most pervasive problems of tropical reservoirs and irrigation systems has been the spread of aquatic nuisance plants. Most early reservoirs were constructed in temperate zone environments where climate, water temperature, lake depth and limited distribution of noxious plants prevented many problems of weed invasion. In tropical and subtropical areas, however, recent reservoir projects, such as at Kariba (Rhodesia and Zambia), Jebel Auliya (Sudan), Brokopondo (Surinam), Nam Pong (Thailand) and Waikato (New Zealand), have suffered problems of rapid infestation.

Aquatic weeds are usually divided into three groups: (a) rooted plants; (b) filamentous algae; and (c) free floating vascular plants. Rooted plants are frequent but usually confined to shallow lakes or shorelines. Filamentous algae are found in most man-made lakes, but tend to be a problem mainly where their rapid growth from sewage pollution and fertilizer runoff causes oxygen depletion. Free floating vascular plants comprise the fewest species, but often cause severe problems, and are the commonest category in tropical regions.

Losses from aquatic weeds usually fall under one or more of the following heads:

1. fishery losses due to competition for light or energy and for nutrients;
2. fishery losses due to the physical interference of weed cover with fishing processes;
3. health losses, because weeds provide good habitat for malarial mosquitoes and disease-carrying snails;
4. possible evapotranspiration losses through increased leaf transpiration, combined with reduced reservoir storage capacity for hydropower and irrigation;
5. recreation losses through interference with fishing and boating;
6. disruption of lake and river navigation; and

7. a variety of losses or damage from weed invasion of irrigation systems, including blockage, water loss, competition for nutrients, decrease of fish and increase of disease organisms.

In some cases, however, aquatic weeds can benefit fisheries by providing shelter and feeding grounds. They may also have considerable potential as a possible source of animal feed for poultry, cattle, pigs and other domestic stock, or of the materials required for producing fertilizer, mulch, paper and packaging material (Lagler, 1969).

Among the water weed infestations that have recently begun to be studied are those caused by the water hyacinth (*Eichhornia crassipes*), a free floating native of South America, now spread widely throughout Africa and Asia. Its populations can build up quickly by vegetative reproduction; in one experiment two plants produced 1,200 in a period of four months. This capacity for growth is well illustrated by the hyacinth's history in the Congo Basin. First reported in 1952, by 1955 it had spread 1,600 kilometres up the Congo river, from Leopoldville to Stanleyville; transportation was blocked, and fish spawning and feeding were interrupted. Control efforts had already required an expenditure of one million dollars by 1957, and the weed was still spreading. Similar histories of infestation have been recorded on the Upper White Nile and in India, Southeast Asia, the Philippines and Central America (Holm *et al.*, 1969).

A particularly controversial question concerns the cost to irrigation and hydropower projects from the possible increased loss of water through leaf evapotranspiration brought about by the invasion of water hyacinth and other weeds. Holm, Weldon and Blackburn (1969) state that: 'The loss of water through evapotranspiration from the leaves has been measured as 3·2 to 3·7 times greater than free evapotranspiration from a surface. This accounted for a loss of more than 6 acre-feet of water in a 6-month period due to a water hyacinth cover. In the dry atmosphere of India, the loss of water through water hyacinth was 7·8 times that of open water (Timmer, Weldon *et al.*, 1967). Even partial coverage of a reservoir by water hyacinth can result in the entire inflow being wasted back into the atmosphere. Thus, the water is not available for hydroelectric power and irrigation.' On the other hand, waterweeds can reduce solar radiation on the water surface and wave action causing evaporation. In addition, many hydrological physicists maintain that the maximum increased water loss possible from surface plant evapotranspiration is about 1·5. Clearly more research is needed in a variety of areas to settle the controversy.

Water lettuce (*Pistia stratiotes*) is another common floating weed that has invaded numerous man-made lakes, such as Lake Volta in Ghana. Unlike the exotic *Eichhornia*, it is an indigenous species. In addition to its possible enhancement of evapotranspiration loss, blockage of navigation and damage to fisheries, this weed poses a significant health threat. The plant is a favourite habitat for the larvae of several mosquito species that carry filariasis and encephalomyelitis. Some of these larvae obtain oxygen from water lettuce roots and never have to surface. In such cases, control of the waterweed is the only effective means of disease control.

Since Surinam's Lake Brokopondo filled in February 1964, a team of scientists has been studying aquatic weed invasion (Leentvaar, 1971). This study is particularly interesting in that a comprehensive study of the ecological consequences of building the dam, published in 1954 by J. P. Schulz, had analysed the potential problem. Water hyacinth (*Eichhornia*) was scarce in the river prior to inundation. By December 1964 it covered 5,000 hectares; by June 1965 17,900 hectares; by April 1966 41,200 hectares (an area equivalent to 53 per cent of the total lake surface). A control programme utilizing 2, 4-D was initiated at an annual cost of $250,000. In addition there has been the cost of losses to water storage from evapotranspiration and fishery impacts, the reservoir fishery having given a poor return in comparison with the previous riverine fishery.

Early planning to anticipate and control this kind of problem emerges as of prime importance wherever tropical or subtropical reservoirs are contemplated. Similarly, the costs of control, or the losses resulting from lack of control, ought to be included in the cost-benefit analyses for man-made lake construction. Planning for control should include surveys of the whole basin in which the proposed reservoir is situated, to identify potential aquatic weed problems. If it proves to be free from pest species, carefully enforced quarantine measures may be necessary to prevent infestation. Local populations should be taught the dangers of aquatic weeds and legislation enacted to cover quarantine and inspection.

Other measures are also possible to help prevent invasion. It may be possible to alter the reservoir height to eliminate inundation of extensive shallows favourable to aquatic plants. Similarly, since lack of clearing of the reservoir area can create good conditions for weed growth, if trees and shrubs emerge above the water surface, complete clearing of the inundation area should help prevent weed infestation, apart from improving fishing opportunities. Control of pollution, particularly by sewage and other nitrogen or phosphate-rich wastes, can do much to lessen the supply of nutrients available for waterweeds. Another technique involves design of the dam itself. Large open spillways will often allow easy discharge of floating weeds. Location of installations away from sites where prevailing winds tend to concentrate floating plants can also help prevent blockage.

Should waterweed invasion occur, several methods of control can be considered: mechanical, chemical and biological. *Mechanical* techniques involve cutting by hand and machine, the use of draglines, and the employment of special control boats. Such techniques, however, are generally impractical for extensive areas, and extremely expensive. 2, 4-D, 2, 4, 5-T, diquat, fenoprop and other *chemical* herbicides have all been used with varying degrees of success. They have proved relatively economical and effective, but may involve toxic disruptions to lake ecosystems and fish life. A serious consequence can be the sudden decomposition of plant matter, causing severe water oxygen depletion and release of toxic substances, such as hydrogen sulphides. Both effects can severely reduce fish production. In addition, crops and forests adjacent to the reservoir, and even irrigated areas downstream from the water body sprayed, may be damaged.

In view of these risks, spraying should probably be limited to areas where neither mechanical nor biological controls are possible. Also, a continuous, low-level spraying operation to hold small weed populations in check may be biologically safer and economically wiser than waiting until infestation has become widespread. *Biological* controls cause fewer secondary problems than chemical herbicides, but little research has as yet gone into the development of a fully effective array of techniques. Habitat manipulation has proven effective in some cases where seasonal control of water level is possible. Rooted weeds are most easily controlled in this fashion, so that drawdown, where shallow areas have high concentrations of free floating weeds, can sometimes be effective.

There are various possible control organisms requiring much more research. For example, the manatee, a herbivorous freshwater mammal inhabiting parts of Africa and Latin America, normally consumes large quantities of aquatic plant material, and has been suggested as one possible control species; initial studies on its potential are currently under way in Florida. Other possibilities include an array of insects, snails, diseases and fish. But not enough is yet known about either the habits and ecology of potential control organisms or the life histories of the aquatic weeds themselves to provide an immediate answer to existing control requirements.

Finally, and perhaps most closely connected with the *mechanical* methods of control mentioned above, the considerable potential of aquatic nuisance plants as sources of animal fodder, agricultural fertilizer and industrial raw material should not be overlooked. In water bodies suffering from eutrophication, water plant harvesting might well help remove the excess nutrients causing oxygen depletion. In Singapore, the Philippines, parts of India and China waterweeds are often grown as a crop to provide fish or pig food (Pirie, 1970). Similar reports indicate potential animal fodder utilization by cattle, buffaloes, poultry and other domestic animals. In Florida, aquatic plant harvesting and concentration into pellet form has produced a high-protein cattle food of some promise. Other productive possibilities include mechanical extraction of leaf protein for use as a human food.

Because of the often high ecological and economic costs of control, such productive potentials may offer a constructive alternative approach that views aquatic weeds as a valuable resource. However, considerable research on possible uses is needed before long-term policies based on the cost of aquatic weed infestations, the cost of controlling them and the opportunities for beneficial utilization, can finally be determined.

(e) Forestry and Watershed Management—Protective Aspects

Forests and their management in river basin developments have a protective value in conserving soil and in maintaining water quality and sustained flow for the purposes of the watershed, irrigation and agriculture. Their value as producers of fuel, timber, special commodities, recreation, wildlife, and for restoring land under shifting agriculture is discussed elsewhere and, in the present context, attention will be confined to the protective aspects directly related to river basin development and summarized in Table 4 below.

Table 4. Tropical watershed management consideration

Environmental factor	Arid watersheds	
	General characteristics	Management consideration
Rainfall	On annual average much less than P.E.T. (potential evapotranspiration). Heavy, intense showers possible.	Control phreatophytes and other vegetation so as to maximize clean runoff but at same time provide protective vegetative cover. Also to stabilize flow.
Soils	Characteristically shallow with high pH, poorly developed structure. Low field capacity for moisture storage. Highly susceptible to erosion.	Terracing and other earth works, as well as retention of vegetative cover to stabilize soil, especially on steeper lands.
Natural vegetation	Xerophytic, sparse, slow growth, low.	Prevent excessive burning and cutting of woody species for charcoal and firewood.
Commonly observed land use	Grazing and browsing by cattle, goats, sheep. Charcoal production. Ephemeral cultivation of short cycle, drought resistant crops. These environments are characteristically deteriorating.	Maintain animal populations at carrying capacity or less. Reduce use intensity where leading to erosion and degradation of vegetation.

Table 4. (contd.) Tropical watershed management consideration

Environmental factor	Sub-humid watersheds General characteristics	Management consideration
Rainfall	Equal to or slightly less than P.E.T. on annual average, but higher than P.E.T. during rainy season, including intense downpours.	Runoff control and regulation measures to maximize both clean runoff to reservoir and availability of soil water for crops and pasturage.
Soils	Variable in structure, depth and fertility, but sometimes with good potential for agriculture. Erosion a potentially serious problem. Rapid runoff may also cause flooding.	Great care required on cultivated land to prevent erosion and rapid runoff, especially on sloping terrain with clay soils.
Natural vegetation	Semideciduous forests, savannah 'parklands' (fire climax).	Can include valuable tropical woods. Natural forest should be left on steep catchment areas.
Commonly observed land use	Short cycle and perennial crops, grazing, forestry. Conflicts can be expected between agricultural and forest uses of lands. High population densities and intensive land use is likely.	Productive agricultural and animals raising uses possible under resource-conservative management. Vigorous soil and water conservation programmes needed in high intensity use, dense population zones. In steep areas natural forests should be conserved, but plantation crops may provide adequate cover.

Environmental factor	Humid watersheds General characteristics	Management consideration
Rainfall	Exceeds P.E.T. in most months. May characterize middle and upper watersheds, subject to orographic influences. Short dry season if any.	Maintenance of natural vegetative cover of paramount importance, in order to absorb rainfall, release clean runoff and stabilize flow.
Soils	Mature soils may be deeply weathered clays with good structure and internal drainage, but low natural fertility. Podzolic soils in cooler zones.	Leaching rapidly depletes exposed soils. Ill-suited for cultivation except on younger fertile soils. Erosion danger variable but slumping and landslides a danger, especially after deforestation.

Table 4. (contd.) Tropical watershed management consideration

Environmental factor	Humid watersheds	
	General characteristics	Management consideration
Natural vegetation	Evergreen forest, including 'rain forest' and 'cloud forest'. Relatively limited commerical value for wood products.	Best left in natural state, especially in steep catchment areas, unless economic perennial crop providing equivalent protection is possible.
Commonly observed land use	Little permanent farming except on exceptionally fertile soils. Shifting agriculture likely.	Soil and forest exploitation technologies not well developed for this environment, except on best soils. Road construction and maintenance costly; fungal diseases serious problem in agriculture.

(*i*) *Watershed Area*

Any change in the watershed area will affect the flow of water and quantity of sediment carried downstream. The protective role of forests in maintaining a relatively stable flow of water for reservoir impoundments and irrigation works is well established. Not only does protective vegetation decrease flood peaks, but it also usually increases water discharge during dry periods. Forests and their root systems are also of great importance in preventing erosion and reducing sediment loads in streams and rivers.

Reservoir projects require particularly close attention to watershed forest conservation. Where upstream catchment zones are subject to deforestation, overgrazing or intensive burning, erosion rates invariably accelerate dramatically; the increased silt loads carried by streams usually lead to more rapid lake sedimentation, loss of storage capacity and reduced reservoir life. In addition, there is very rapid runoff after periods of rainfall, with the result that, if lake storage capacity is already fully utilized, much of this water passes over the spillway and is lost for future power and irrigation needs, but may subject valuable lands to damage by flooding. During dry spells, poorly managed watersheds invariably release less water to downstream areas since less has been retained following precipitation. This can adversely affect flows vital to lake and irrigation works. It is therefore important that reservoir projects provide for the survey and planned management of the forests of the catchment zone.

Of equal importance for the ecologically sound management of catchment waters is attention to road building and casual settlement. In the absence of proper resettlement programmes, inundation of the reservoir site is likely to be followed by spontaneous colonization of watershed lands, which can lead to serious erosion and the silting of the lake and irrigation channels. Ill-considered construction of new roads, trails and other facilities in the catchment can likewise act as a spur to unplanned settlement, with identical results.

Depletion of storage capacity tends to be greatest in smaller reservoirs in which the highest annual storage loss rates are normally found. In larger and/or deeper reservoirs with generally lower loss rates, the situation varies widely according to the silt burdens of inflowing streams and rivers. Sedimentation not only affects available storage capacity, but can also be detrimental to fishing, ecosystem production processes, navigation, chemical quality of the water, and recreation. Exclusion of silt from water passing downstream of a dam may also have adverse consequences, such as the undermining of bridge abutments, accelerating channel erosion, and the disruption of geomorphological equilibria in delta zones. An instance of the latter has recently been reported in the Nile Delta in consequence of the effects of the Aswan and other dams in reducing silt deposition (Kassas, 1972).

In most reservoir and irrigation projects, the problems of sedimentation and achieving sustained stream inflow can usually best be overcome by careful protection and management of upstream forests and grasslands. Commonly, the natural vegetation cover constitutes a more stable ecosystem than artificial forest plantations, which tend to be composed of relatively few exotic species, provide less comprehensive cover and, therefore, less protection from erosion. In addition, stands of one or two species are much more susceptible to plant disease; a serious infection may mean the partial destruction of watershed cover. In some cases, however, particularly in arid zones where its permanent establishment is feasible, a grass cover in the catchment can be used to reduce evapotranspiration losses. In less arid areas where evapotranspiration is less serious than erosion or uncertain stream-flow, afforestation may best rehabilitate catchments degraded by destructive human impacts. Unstable soils along stream banks, the erosion of which by flash floods is a frequent contributory factor in heavy silt loads, call for special revegetation efforts.

Forests are a characteristic feature of mountainous areas with relatively high precipitation and steep, erodable soils. Their conservation and protection in such sites is particularly crucial. In the western USA, comparison of burned versus unburned watershed forests showed that storm runoff increased up to 100 times following fire, and erosion by 50 times (United Nations, 1970). Nevertheless, it should be recognized that forest areas exist under a wide variety of hydrologic situations, slopes, soils and human pressures. The structure and function of forest ecosystems also vary widely and differing management objectives may change one's perspective. Thus, in some areas maximum reduction of evapotranspiration may be sought; in others, flood and erosion control may be the primary aim, in still others timber production. All these factors must be considered when surveying watershed functioning and determining management policy. Every area will usually require highly individual analysis.

Grassland, either natural or man-induced, is usual in areas where water supplies are critical. Many grasses tend to transpire less water than forests and are usually well suited for semi-arid watersheds. Others, however, like the long-rooted Kikuyu grass, may transpire more water than certain species of trees. In short, grassland ecosystems are as variable as forests and subject to widely differing natural and human impacts. Thus the shallower-rooted species usually use less water but, in arid areas, deep-rooted species may be better adapted to survive. Similarly, different grasses utilize water more or less efficiently. Some require more water but produce great quantities of forage; others require less water but are much less efficient in producing grass. Where other uses in addition to watershed protection are contemplated, these differences may be important. Just as in the case of catchment forests, grassland areas may be needed and used for a variety of purposes. They may primarily be required for reducing evapotranspiration losses; they may otherwise contribute to checking water losses and flash flood erosion; or they may be needed not only for these purposes but also for grazing. In the latter event, special care is needed since grazing invariably increases both erosion and runoff; in excess it may also be a cause of pollution.

Another common type of watershed land-use is cultivation. Numerous techniques are available to reduce both erosion and water loss from croplands, particularly on sloping soils. Indeed, badly-managed agricultural areas are often primarily responsible for watershed erosion and water loss problems. Contour ploughing, terracing, planting of cover crops, strip cropping and careful soil conservation survey and extension services, can help to improve degraded watershed land under cultivation or, more importantly, to prevent degradation before it occurs.

In addition to their largely protective functions, forests and grasslands of catchment areas can provide productive multiple benefits from forest products, wildlife and recreation opportunities, as well as grazing. In some more vulnerable watersheds, however, intensive logging and grazing may be incompatible with adequate erosion control. Management planning should be careful to develop timber or grazing production programmes that do not conflict with basic protective functions of the vegetative cover. Large-scale clearing, burning, monocultural plantation, herbicide and insecticide use, or road building may all have destructive downstream impacts. On the other hand, park recreation, natural area research, fishing and wildlife utilization are very unlikely to conflict with protective functions.

(ii) Shoreline Area

Reservoir construction often alters ground-water patterns around the shoreline. In some cases, forest growth will be assisted, in others (particularly lakes whose level fluctuates considerably) it may be inhibited. Normally, indigenous species adapted to riverbank conditions will do less well along reservoir shores, due to more static water conditions or irregularity of water levels. For this reason new species, tolerant of such conditions, may have to be introduced. They need to be well-tested, together with surveys of climate, soil and other environmental parameters, to determine the most suitable species. Care should also be taken not to introduce exotic species that might invade adjacent areas and become pests.

The benefits of sound forest management along shorelines can be considerable. Forests can serve as windbreaks, stabilizers of erosion-prone soil, barriers to overgrazing and settlement, areas for recreation and wildlife, and as sources of various products. On the negative side, however, some trees, such as tamarisk, transpire great quantities of water; this may represent a serious loss in arid zones where water supplies are critical and may also concentrate salts in surface soils. Lagler (1969) comments on the subject as follows:

'Trees transpire large amounts of water owing to their extensive root systems and their dark foliage which may absorb greater amounts of solar energy than light-coloured surfaces. Whereas this comparatively high transpiration rate varies with the local climate, a forest is likely to remove more moisture from the soil than a less tall form of cover, such as grass. Thus, where drainage is important, tree plantings may be advantageous, but where soil moisture must be conserved, a less dense, shallow-rooted plant cover may be required. Of course, the protective function of forest cover must always be weighed against its water-using function, and may be of overriding importance.

'Some water-loving tree and shrub species can create special problems on particular sites. On flood plains economically useless species with very deep root systems may invade land which then becomes unavailable for other purposes such as grazing. If the water table is high, the replacement of these plants by pasture may be relatively simple. If it is low, however, the establishment of useful plants can be difficult unless irrigation is possible.'

(iii) Inundation Site and Downstream Area

An important cost often neglected in man-made lake projects is the direct loss of forest land flooded by the reservoir. Good soil moisture and nutrient availability often makes such forest productive. In any attempt to define ecologically realistic cost-benefit ratios for reservoir projects, the total value of the forest resources lost through inundation should be included. Similarly, the expense of clearing lanes or larger areas of standing vegetation within the reservoir site ought to be included as a cost factor, if investigation shows that clearing is necessary, as it often is, to prevent waterweed invasion, interference with lake navigation and obstacles to fishing. Where forests are not cleared from the site, processes of decay after flooding tend to be slow. Thus, trees submerged in 1903 under Panama's Gatun Lake remain standing today (Challinor, 1969). Wherever possible, clearing of the reservoir site should be linked with inventory and utilization of valuable timber species that might otherwise be lost.

Areas downstream from a dam may contain forests adapted to periodic flooding and particularly to nutrients deposited by flood waters and the effect of flooding on soil structure. Completion of the dam may alter the regime unfavourably for such species. The possibility of remedying the situation by application of fertilizers or by introducing better adapted species should be investigated. Other positive opportunities may exist in downstream areas, particularly for irrigated forest production. Based on the findings of appropriate soil and botanical surveys, the eventual emphasis may be on plantation forestry or on incorporating some forested sectors in the irrigation system supplied from the reservoir.

Forest protection or planting may be particularly necessary along downstream river banks subjected to higher rates of erosion because of the deposition of most of the silt-load in the reservoir. It may also sometimes be usefully applied to the protection of deltas threatened by increased marine erosion due to the same causes.

(f) Recreation, Tourism and Protected Areas

The general considerations affecting the development of tourism and National Parks, and particularly their ecological impacts in coastal and island environments, have been discussed in Chapter 5. However, the importance in this context of man-made lakes and their rather unusual mixture of positive and negative relationships with recreation and tourism merit some further comment.

(*i*) *Recreational Benefits*

If carefully designed and managed, the use of a lake as the base for recreation and tourism, comprising such activities as boating, swimming, water skiing, fishing, hunting, wildlife observation and photography, touring and camping, can be beneficial to the region in which the lake is located. Any reservoir development should, therefore, include an adequate survey of and conservation plan for the protection, restoration and enhancement of outdoor recreational resources. Areas suitable for setting aside as wildlife reserves (particularly in marshes and on islands), parks or forest reserves, and any historic and archaeological sites, in or near the reservoir environment, should be identified. Their potential for tourism and recreation, as well as their value as catchment protection, fish spawning or feeding grounds, and so on, can then be considered and plans for legislative protection, provision of guards, management needs, trained administration and funding should form an integral part of the overall plan for the reservoir and its management.

The forests of a catchment area, in addition to their several functions discussed in the previous subsection (e) of this chapter, can have many other resource values, such as scientific reserves, wildlife conservation areas, wilderness zones, hunting areas, and centres for recreation and tourism. Some of these potential uses may conflict—the establishment of national parks, for example, might preclude public hunting—but others can be compatible with national park status, such as recreation, gene pool preservation, erosion control and wildlife protection. Comprehensive resource capability surveys will be needed to develop an integrated approach to selecting and designating which areas are best suited for specific purposes. Compatible and incompatible uses should be identified in relation to each proposed unit in the plan, and management techniques developed that are appropriate to each unit as well as overall. It is particularly important to locate unique and irreplaceable natural areas, scenic values and major historic or archaeölogical sites early in the planning process, since these can so easily be lost through conflict with other uses, and therefore, whenever possible, must be given protection before development is initiated.

(*ii*) *Costs to Recreation and Science*

Although reservoirs offer opportunities for lake-based recreation and tourism, they also often inflict major losses when they inundate unique natural areas, existing or potential parks and wildlife reserves, or important historical and archaeological sites. A realistic cost-benefit appraisal of a projected reservoir should therefore take account of the full cost of such losses in terms of recreational, scientific, cultural and touristic value (Figure 30). Even more importantly, these values should be assessed early enough in the prefeasibility stage of a reservoir project to allow serious consideration of alternative sites.

Reservoir inundation often destroys natural features of free flowing rivers. Waterfalls, rapids, canyons, wildlife and rare plant communities, may be permanently lost through flooding. In many cases, the potential of these resources for park and reserve-based tourism and recreation, and the loss to science represented by the destruction of a complex flora and fauna, are very considerable, especially if rare or endangered species and habitats, found nowhere else, are involved.

Figure 30. Before water development is undertaken the full extent of recreational, scientific, cultural, and touristic values to be lost need be considered. Blue Nile, Ethiopia. (Photogaph by F. Vollmar: courtesy World Wildlife Fund).

Scientific studies in parks and reserves can have great value in helping towards an understanding of ecosystem functioning. Most tropical river basins are extremely complex environments about which relatively little is known. Basic information on the life cycles of beneficial soil organisms, insects, fish, wildlife and plants is still often lacking. Even simple inventories of potentially useful plants and animals are rarely undertaken prior to reservoir inundation and other river basin development. Similarly, very little may yet be known about the ecology of certain disease organisms and pests, as they function in natural environments. If no attempt is made to learn the values of the complexity of species likely to be affected by the development project, losses to science may well become costs to society as a whole and to governmental budgets in particular.

The value of native plants and animals as a local food and material resource can be quite important to human populations living in the reservoir region. Surveys of these biotic resources, utilized by such communities, and the potential impact on them of a man-made lake, should be included at an early stage of planning and, wherever possible, measures for their salvage, restoration and management should be built into the project. These measures will of course be critical in cases where native flora and fauna form a really important part of the diet of the local people.

(iii) Summarized Recommendations

River basin projects should include—

1. Survey and inventory of areas which might qualify for preservation due to important scenic, watershed conservation, biotic, scientific or historic resources.

2. Determination of costs due to inundation loss and/or needed to cover the salvage, study and restoration of the aforementioned resources.

3. A comprehensive plan for adequate protection and management of non-inundated areas deserving protection (including modification of other related basin planning, such as road development). Identification of the compatible and non-compatible uses of each specific site ought to be included.

4. Provision for recreational and touristic access, accommodation and information programmes, where compatible with resource protection objectives.

5. Inclusion of all the costs involved in the preceding recommendations within cost-benefit analyses at the stage preceding feasibility studies or even earlier, when the criteria for choosing the project site are under examination.

6. Consideration of long-term research in ecosystems relevant to the project, including possible establishment of environmental monitoring centres. The creation and effective protection of scientific reserves, covering a representative sample of the region's significant ecosystem types, is an essential corollary of research, whether the latter is basic (investigating biological processes) or applied to human resource needs: provision for both types of research is equally important.

(g) Earthquakes and Reservoirs

One possible consequence of filling a reservoir which has received only rather limited emphasis, is increased liability to seismic activity. Numerous studies have documented this phenomenon in detail and need not be fully examined here. For example, associated seismic activities have been reported from the region of the Boulder Dam (USA), Koyna (India), Kariba (Rhodesia), San Luis (USA), Vajont (Italy), Mongla (Pakistan), Catalogne (Spain), L'Oud Fodda (Algeria), Kremesta and Marathon (Greece) and Contra (Switzerland). Earthquakes registered as high as 6 on the Richter scale are recorded in the case of Koyna, Kariba and Kremesta.

Current research indicates that the height of water in the reservoir lake may be more important than total volume. Often the fore-shocks are less powerful than later shocks (unlike many natural earthquakes). Special geological conditions are usually responsible for such seismic activity and research shows that areas typified by diaclases with water-leakage, zones of faults and fault mosaics, and heterogeneous underground strata, may all be prone to tremors from filling man-made lakes (Rothé, 1971). Other related factors include: underground water penetration rates, water pressure and rock lubrication, the activity of rock pore air, creation of artificial faults, and presence of natural faulting.

Because of the possibility of serious earthquakes near man-made lakes following filling, it is obviously important to undertake detailed geological surveys of the reservoir site and surrounding region very early in the feasibility study stage (Rothé, 1971). The results of disregarding potential disruption of geological factors can be particularly severe where substantial human settlements exists in a susceptible reservoir region. The possibility that earthquakes might result in the destruction of a dam threatens catastrophe to populations living downstream from the reservoir and dam site.

(h) Human Problems Associated with Reservoir Impoundment and Resettlement

One of the most difficult and potentially costly aspects of man-made lake development is the inundation of inhabited areas, and resettlement (both planned and spontaneous) of displaced populations. Many reservoirs in the tropics have flooded out substantial numbers of people from their ancestral homes and land, forced resettlement in poor and environmentally different sites, caused grave disease and nutritional problems, and generally disrupted the social fabric.

In view of the costs of resettlement, thorough social and environmental investigations of communities facing displacement are needed very early in the prefeasibility phase of man-made lake projects. Such studies should, as a minimum, cover public health, nutrition, land use, social structure, population factors, regional administration, educational systems and housing.

In addition to studies of the population in the inundation area, possible social, ecological and economic difficulties that might be encountered in the resettlement zone deserve particularly intensive analysis. Alternatives to planned rural resettlement sites—such as urban immigration and employment, compensation sufficient to allow evacuees to purchase good land elsewhere, and absorption into fishing, irrigation, agriculture and other activities related to the reservoir—should be investigated. Finally, a complete account of the full social and ecological costs of resettlement (including alternatives) should be included in determining the cost-benefit ratio that measures the advisability of the project.

The following checklist covers the costs which are commonly incurred as a result of people having to move from the impoundment area of a dam. These are all problems which have often been involved in past relocation programmes and which should be anticipated and assessed as accurately as possible in future resettlement planning.

1. Loss of agricultural, forest or minerally productive lands.
2. Loss of homes, villages, religious and cultural sites.
3. Loss of other community facilities (schools, health centres, etc.).
4. Resource difficulties related to break-up of social units and families.
5. Inadequate notification to evacuees prior to inundation, and misleading information on resettlement sites and facilities.
6. Resettlement site selection without studies of soil conditions, agricultural potential, water needs and supplies, disease or transportation requirements.
7. Political constraints on lands otherwise potentially available for consideration as resettlement sites.
8. Tendency towards selecting government forest reserve lands for resettlement, leading to loss of valuable watershed protection and timber production.
9. Inadequate and hasty studies of the particular characteristics and needs of people to be resettled, and programmes for their retraining: a very common flaw has been lack of opportunity for evacuees to express their needs and desires to the planners responsible.
10. Lack of funds, planning capability and trained personnel to implement and manage a successful resettlement programme.
11. Lack of funds, planning capability and personnel to retrain evacuees and educate them in the needs, dangers and opportunities of their new environment.
12. Difficulties involved in having to move people from fertile lowland agricultural areas to poorer quality upland sites, including—
    (i) lower agricultural production associated with less productive soils and unfamiliar land use practices;
    (ii) exposure to new disease environments and intensification of existing diseases, particularly following a rise in population density;
    (iii) reduced crop diversity and loss of river fisheries, contributing to nutritional impoverishment;
    (iv) lack of potable household and irrigation water; and
    (v) fewer potential cash crops and loss of income.
13. Inadequate compensation of evacuees for losses, sometimes based on faulty land valuation.
14. Spontaneous colonization of areas in the reservoir catchment zone causing accelerated erosion and lake sedimentation.
15. Competition for land occupied by previous settlers—usually upland

shifting agriculturalists or lowland farmers—with resulting pressure on carrying capacity.

16. Competition for cultivable land and for reservoir fishing opportunities from immigrants of more distant origin, who may be in a stronger position to launch new enterprises.
17. Failure to develop all possible means of enabling evacuees to share in the benefits from the reservoir.
18. Inadequate attention to population growth rates in resettlement areas and to providing birth control advice and assistance.
19. Lack of research on possible new fisheries, livestock or crops which might be useful on resettlement sites, particularly in uplands.
20. Failure by planners and responsible agencies to investigate, fund and develop a range of viable options on which the people to be resettled are consulted and from which they may choose, such as, for example—
    (i) relocation and retraining in urban areas;
    (ii) full-time fishing and aquaculture;
    (iii) relocation in new irrigated areas (if any) served by the reservoir;
    (iv) sufficient compensation, on an individual basis, to cover purchase of good land elsewhere; or
    (v) sufficient compensation to cover such purchase communally.

It is becoming apparent that resettlement is often the largest and most complex issue related to man-made lake development. Indeed, it is often the very size and complexity of the issue which causes planners and executors of such projects to neglect or ignore it. Nevertheless, particularly where large communities will be affected, all the above-listed factors must be taken into account if project planning is to be ecologically sound and socially equitable.

In addition to the problems of resettling populations evicted by the filling of a reservoir, a further problem may arise when downstream populations are adversely affected by altered water regimes, loss of soil nutrients from seasonal flooding and increased channel erosion. In some cases, these impacts have been severe enough to force resettlement.

Thayer Scudder (1972), in his analysis of the Kariba Dam, investigated downstream impacts along the Zambesi River. Prior to the dam, a tribal farming population of at least several thousand lived close to the river; they produced two crops (largely subsistence crops) annually of legumes, cereals and cucurbits. The first crop was planted at the start of the rainy season on deltas and riverine alluvials; it was harvested prior to the river flooding. A second crop was sown as flood waters receded. Following construction of the dam, river flow below it became extremely irregular.

For three years, there was no normal flood flow. Thereafter, flooding occurred intermittently, but was not synchronized with downstream agricultural practices. During the low-water period, areas formerly available for dry season cropping following the normal seasonal flood had to be abandoned due to the unpredictability of flows. Later on, many crops planted on previously unused flood plains were destroyed by irregular flooding. Since then, intermittent releases of high and low quantities of water, uncorrelated with former natural discharge patterns, have made traditional agriculture in the downstream basin difficult and impractical. Scudder concludes: 'I am not aware that those planning for Kariba even considered alternate outflows for the future development of the downriver area. Nor am I aware that they considered the costs of possible food shortages arising from the present regime. . . . This project was essentially a uni-purpose scheme.'

Many examples exist where, through the neglect of one factor or another, resettlement projects have failed. Likewise, sound ecological advice has sometimes failed in its application through ignorance of cultural restrictions on its acceptability. At present, numerous studies for new man-made lakes are under way in Africa, Asia, Latin America and the Pacific region. In the recent past, 80,000 people were resettled as a result of Ghana's Volta Dam project; in Egypt and the Sudan, approximately 120,000 people were similarly affected by the Aswan High Dam; in Nigeria, 42,500 people by the Kainji Dam; and in Thailand, 25–30,000 people were displaced at Nam Pong, and 30,000 at Lam Pao. Estimates based on aerial surveys of the projected impoundment area of Laos and Thailand's Pa Mong dam, show that approximately 310,000 people are at present living in the area. Population growth is at a rate of 3 per cent per annum, so that the number of people affected eventually, when inundation occurs, could be considerably higher.

One of the most important aspects of resettlement planning is its timing. In many past cases, little attention was given to resettlement until actual preparations for dam construction were under way (Scudder, 1966). At such a late stage, it is often extremely difficult to undertake adequate census and population distribution surveys, studies of land use, soil capability, nutrition and culture, ecological surveys, and intensive analysis of resettlement sites or alternative resettlement options.

Unless the necessary studies (economic, social and ecological) are initiated at the earliest possible stage of planning of man-made lakes, resettlement becomes a crash programme that inevitably involves serious planning oversights. The data for formulating realistic cost-benefit analyses of the project's feasibility are simply no longer available.

Because of the substantial economic, environmental and social problems and costs involved in most sound resettlement programmes, a special effort should be made to resettle people in areas of similar soil capabilities, land use potentials and environmental constraints as those they are leaving. Such sites should ideally be as close to their former home as possible and not already occupied by others. If this is not feasible, the basic objective in selecting the site for resettlement should nevertheless always be to make the people concerned beneficiaries rather than victims of development.

When a population can be relocated at suitable population densities in a similar ecological situation, the process of environmental and social adaptation is usually much easier, has a higher chance of success, and is less costly than relocation in substantially new ecosystems. Social systems will normally be well adapted, as will land use traditions, crop patterns, disease control patterns and balanced nutritional knowledge.

Unfortunately, relocation in closely identical environments is often not possible. Displaced populations are usually farmers and/or fishermen in river edge and alluvial areas which commonly have relatively high productive capabilities. Particularly where man-made lakes are constructed in densely populated areas, most of the desirable alluvial land not inundated is already intensively settled and used, and will not be available for resettlement purposes. Over much of the world, such situations tend to be the rule, not the exception. When populations displaced by reservoirs are of considerable magnitude, finding suitable relocation sites (of sufficient size) on ecologically similar land not already occupied may pose particularly difficult problems. For these reasons, it has been and will continue to be necessary to consider the other solutions mentioned under item (20) of the checklist, namely—(a) reservoir fishing and aquaculture; (b) upland agriculture and grazing; (c) relocation on newly irrigated lands; (d) compensation sufficient to allow purchase of prime land elsewhere; or (e) migration to urban areas.

In most projects, reservoir fishing and aquaculture have at least some potential for absorbing part of the displaced populations. However, past experience indicates that fishing and other lake-shore activities are often taken over by people from outside the project region and, secondly, that many agricultural or pastoral peoples evicted by flooding may have difficulty in making the social transition to new occupations (particularly where training and services are lacking). Similarly, where new lands are brought under irrigation, additional people may be able to change over to this form of agriculture but, in both cases, large investments on compensation, training and services will be needed. Often, however, neither of these solutions may be feasible, for cultural or ecological reasons.

It is possible that in some instances enough compensation might be provided to allow individuals to purchase good land elsewhere, construct homes and enjoy other social facilities. But frequently there may not be enough good land available to attempt such a programme. Many other difficulties also face individuals attempting to start a new life, such as the inevitable break up of family and other social units; political constraints; unfamiliarity with new cultures and new land use practices; inadequate funding and retraining; and the fear of moving to unknown areas.

Finally, what of the alternative solution of allowing or encouraging people to move into urban areas? Where sufficient funding and skilled personnel are made available to provide for intensive re-education and training for the new skills required by urban living, where urban centres have a shortage of unskilled or partially skilled labour, and when adequate compensation is made to displaced individuals to cover adequate housing, health care and other services needed to enable them to make an effective start on urban life, this alternative may be viable. Unfortunately, these basic requirements often simply cannot be met where agencies responsible for evacuees are lacking in funds and experience, where the cities and towns already swarm with the unskilled and unemployed, and where adequate compensation is not economically feasible.

For all the above reasons, the alternative of upland resettlement is the one chosen in the majority of relocation projects initiated to date. Although many populations cultivating or fishing alluvial and riverine areas commonly indulge in some additional shifting cultivation on upland sites, a complete change over to an upland land use pattern usually generates a series of serious consequences. In addition to involving loss of agricultural and forest lands and of village facilities, and the break-up of social units, many upland sites are not suitable for intensive agriculture under the dense populations that can be sustained by lowland alluvial soils enriched by regular flooding, such as those of the rich wet rice areas of Asia. Once cleared, tropical upland soils are likely to suffer rapid fertility loss through nutrient leaching and erosion. In addition, irrigation water is rarely available and fish protein is no longer in immediate supply.

The evacuees' problems may be aggravated by: cultural unfamiliarity with techniques for intensive upland agricultural production; reduced crop diversity (both subsistence and cash crops); nutritional impoverishment and loss of cash income; exposure to new diseases; lack of potable household water; and possible competition with groups of shifting agriculturalists who may have traditionally utilized such upland sites under less intensive land use regimes. These difficulties may be made still worse if population growth places increasing pressure on the social and environmental systems of the relocation site. The cumulative impact of such problems may lead to severe degradation of both the human population and land resources of upland resettlements, as shown in Table 5 below. There are, however, certain soil types, particularly those derived from recent volcanic materials and those undergoing rapid enough erosion to ensure a continuous input of minerals, which may be more favourable for upland agriculture.

Table 5. Summary of problems observed in resettlement of populations from reservoir impoundment areas

Environment and resource problems	Social and human problems
1. Land available for upland resettlement less productive than riverine and floodplain soils. If soil management not adequate, decline in fertility and erosion may occur.	1. Population is exposed to new diseases to which they have little or no resistance.
2. Resettlement sites often productive forest land, which is thus converted to less productive agricultural uses.	2. Resettled population is in competition with previously established populations (usually shifting agriculturalists) in resettlement area; land tenure problems.
3. Spontaneous resettlement in steep catchments leads to deforestation, resulting in erosion and runoff control problems.	3. New land tenure and ownership patterns among resettled population results in social strains, possibly eroding community ties. Families and extended families may be broken up.
4. Resettlement area is a new disease environment.	4. Dietary shift, resulting from new food crops; possible imbalance, leading to malnutrition.
	5. If resettlement is to urban areas, multiple problems of social, cultural and economic adjustment; possible high social costs.
	6. General disruption of human ecology and man/land relationships.

Production problems	Management and administrative problems
1. Settlers unfamiliar with production techniques needed for new crops and different climatic and soil conditions.	1. Presettlement surveys and studies inadequate.
2. In uplands, production technology based on flood-plain farming is unsuitable: temporary drop in production virtually inevitable, and may be prolonged.	2. Studies made but no follow-up in action programmes.
	3. Inadequate notification and orientation of affected population.
	4. Inadequate compensation to families evacuated but not resettled.
	5. Indifference to response of affected population; no consideration of their felt needs.

Because of the complex social and ecological processes involved in upland resettlement, the necessity very early in the planning process for careful studies of soil capability, potential crop patterns, cultural acceptability and extension requirements, is again emphasized. In addition, funds to establish local upland agricultural and grazing experimental stations ought to be included in projects involving major resettlement. These stations could be invaluable in carrying out research on socially, ecologically and economically sound crop patterns. They could also undertake pilot programmes, well in advance of full-scale resettlement, in order to work out land use patterns that are socially feasible as well as economically and nutritionally sound.

Most importantly, experience from other areas should be drawn upon to help improve planning and the training of resettlement personnel; this would assist in ensuring successful development and take advantage of many positive opportunities for relocation that until now have often been disregarded. Another major failure of most planning has been a tendency not to seek or consider the opinions of the people being resettled. Consultation of these opinions should form an integral part of all future resettlement work.

To sum up, development projects for reservoirs involving the resettlement of populations should always aim at improving the living standards of all the people of a river basin. With this aim in view, the design of integrated and comprehensive preinvestment studies must take full account of the wishes and development ambitions of the total population, particularly those forced to move from their original homes and lands. If planning, background studies and extension services receive the financial and other support they deserve, it could result in substantial benefits for those resettled as well as for overall river basin development.

(5) IRRIGATION PROJECTS

(a) Introduction

Irrigation technology is an important key to major increases in agricultural production in developing countries. Under proper management, irrigated crops can produce very high yields. In particular, when such crops can be grown in the drier atmosphere of arid zones, they are relatively free from the numerous fungal and bacterial plant diseases that thrive in the humid air of wetter zones, and, in general, hold out a greater promise of success. Water availability is of course the main limiting factor. For all these reasons, large investments are being made on irrigation projects in arid zones with arable soils.

Despite the fact that irrigation technology is one of the oldest agricultural techniques, certain problems tend to emerge repeatedly. One of the commonest is salinization which, along with sedimentation, is believed to have brought about the ancient decline of agriculture in Mesopotamia (Jacobsen and Adams, 1958) and continues to jeopardize the success of many modern irrigation projects. A more recent complication stems from the use of modern agricultural chemicals for fertilizing and for pest control, which can contaminate irrigation water. Both salinization and water contamination are frequently associated with drainage difficulties. These impacts sometimes represent greater engineering problems than those involved in providing the necessary supply of water.

The agroecosystem under irrigation probably receives more intensive management than any other rural ecosystem occupied by man. In arid zones, especially, the successful growth of crops through irrigation is almost totally dependent upon man's ability to manage the soil–plant–water ecology, so that optimum quantities of water are supplied to the plant at the correct time. Although a great body of scientific and technical knowledge has been amassed on crop growth under irrigation, a common weakness of many projects is the apparent failure to apply this knowledge. Application of too much water is a particularly common failing.

These difficulties are dealt with in some detail in the following pages. Water-borne diseases, waterweeds and resettlement questions, which have already been discussed in earlier sections of this chapter, also have an obvious significance for irrigation schemes.

(b) The Agroecosystem under Irrigation

The characteristics of certain ecosystems in relation to development planning have been reviewed in Chapters 3 and 4. However, it is worth reiterating that the sub-humid or arid ecosystem is inherently unstable. This is illustrated by the potential for sudden and large increases in insect and (in some regions) bird populations, triggered by the appearance of an extensively cultivated food source. Irrigated crops represent one such source. This is why large irrigation projects should always have on their staff technicians experienced in pest control, to monitor changes in populations of potential plant or animal pests and coordinate control programmes. For the reasons discussed in earlier chapters, these programmes should always aim at minimal use of chemical pesticides and herbicides.

(c) Salinization and Drainage

Irrigation project planning should anticipate the salinization of soils under irrigation and invasion by salt-concentrating plants. Engineering as well as management techniques should be built into a project in order to forestall the possibility of losing land to salinization. Solutions have had to be sought for the problem of reclaiming salinized soils in several large irrigated areas in the world, notably the Imperial Valley in California and the Punjab in Pakistan, both of which were seriously affected by salinization directly associated with irrigation. Economical means of de-salinizing water have yet to be developed, so the correction or prevention of salinization has had to be accomplished through water management techniques adjusted to local climatic and geohydrological conditions. The principal reclamation techniques are flushing, pumping to lower the water table, and careful regulation of irrigation flow.

The task of the planner is to ensure the collection of sufficient data at the preinvestment survey stage to enable the potential for salinization to be predicted and to provide for appropriate structures and controls on water, soil and crop management which will minimize that potential. The high cost of reclamation of salinized soils, particularly those affected by sodium compounds, fully justifies the early consideration of preventive measures.

To keep irrigated soils free of toxic levels of salt, sufficient water to flush out the salts must be applied. However, the fate of the salt-charged runoff should also be a major consideration. It may leach into underlying aquifers and increase the salinity of ground water as well as raising the water table. If the water table is raised to within a few feet of the surface, drainage of irrigation water is impeded and salinization will be difficult or impossible to prevent; if it is raised even closer to the surface (usually 3 feet or less), the ground water will evaporate and contribute to salinization. In certain irrigated areas of the Helmand Valley of Afghanistan, impeded subsurface drainage caused the water table to rise, with resulting salinization. An expensive reclamation programme, involving the pumping of ground water to lower the water table, had to be undertaken (Michel, 1972). The introduction of perennial irrigation in the Punjab region of the Indus Valley, Pakistan, aggravated the area's naturally poor subsurface drainage and raised the water table to within a few feet of the surface. By 1960, 8 million of the 13·7 million acres under irrigation were affected by salinity and 3·5 million acres went out of production. To correct this situation, thousands of costly tube wells have had to be constructed to serve the dual purpose of supplying irrigation water—where ground water is of good quality—and to lower the water table (Greenman *et al.*, 1967). Although some economic plants tolerate relatively high concentrations of salt (as shown in Table 6), saline runoff still constitutes a problem if it is required for reuse or is discharged into fresh water bodies with economically important aquatic life.

Table 6. Relative salt tolerance of selected vegetables, field and forage crops[a]

	Vegetable crops	Field crops	Forage crops
Greater tolerance to	Garden beets	Barley	Saltgrass
	Spinach	Sugar-beet	Bermuda grass
	Tomato	Cotton	Fescue
	Broccoli	Rye	Western wheat-grass
	Cabbage	Wheat	Barley (hay)
	Sweet corn	Oats	Birdsfoot trefoil
	Potato (white rose)	Rice	Yellow sweet-clover
	Carrot	Sorghum	Mountain brome
	Peas	Field corn	Sudan grass
	Squash	Flax	Alfalfa (California common)
	Cucumber	Sunflower	Meadow fescue
↓	Radish	Castor beans	Smooth brome
Lower tolerance	Green beans	Field beans	Clovers (alsike, red and ladine)

[a] Data from Report of the Committee on Water Criteria, 1968, p. 150. Copyright Conservation Foundation, Washington D.C., USA.

The two major considerations to be taken into account by the planner with respect to the potential for salinization are, therefore, the drainage characteristics of the area to be irrigated, and the downstream uses of water receiving salt-charged runoff from irrigated fields. Preinvestment surveys should include appropriate hydro-geological investigation of the ground-water table and estimates of potential subsurface drainage problems that might follow inauguration of the project. Such surveys may indicate the desirability of a network of wells to monitor changes in the water table and in the water quality of the aquifer. Where the contamination of field runoff with salts is inevitable, however, an assessment should be made of how this will affect the subsequent use of this water as it flows into streams or percolates downward into aquifers. It may be possible to direct saline runoff so as to minimize its polluting effects and, where salt contamination is likely to be serious, this needs investigation. In the operating phase of the project, provision must be made for monitoring salt levels in the irrigation water.

(d) Irrigation Water Contamination

Contamination of water by pesticides, salts, and other pollutants is an aspect of irrigation that should be considered at the planning stage, not only to determine what engineering solutions might be applied to reduce it, but also to provide for the technical services required to monitor contamination levels and coordinate the maintenance of water quality. This is of particular importance in countries where field and canal water may be used not only for irrigation, but also for household uses and bathing, as well as for raising fish and domestic waterfowl.

The insecticide lindane, used to control rice stem-borer, has proved to be lethal to tilapia, the species of fish commonly raised in Southeast Asian rice fields (Kok, 1972). Dalapon, a waterweed herbicide employed in canals and ponds, is a skin irritant and could harm people bathing in waters contaminated with it. The herbicide sodium arsenite is extremely toxic not only to most plants, but also to all other life forms. It should obviously never be applied where it can contaminate water used for bathing or domestic use, or water supplied to rice-fields (rice is extremely sensitive to arsenic).

The planning of an irrigation project should, therefore, anticipate contamination problems associated with domestic and other use of water supplied to the fields, and also of canals and streams receiving return water from the project. Where serious water quality problems can be expected, it may be necessary to provide alternative domestic water supplies for the people concerned, who in rural areas may well be accustomed to using untreated water for drinking, cooking and bathing. It may also be necessary to devise measures for keeping water used for raising fish and waterfowl free of toxic chemicals. These problems need special attention in arid regions with poor or non-existent potable water sources, particularly when wells or water treatment plants are being considered.

Problems of water pollution by pesticides and herbicides are discussed in subsections (d) and (g) in part (2) of the previous chapter; the main point to add in the present context is that herbicide leachates could contaminate ground water and present complications to projects which pump water from this source for irrigation. Leaching depends upon soil characteristics, rainfall, surface and subsurface drainage and rate of pesticide application, and varies according to type of herbicide. Generally speaking, clay and organic matter absorb and thus immobilize or deactivate these chemicals (and also some insecticides), while leaching is greater in sandy soils. Because degradation and breakdown into less toxic compounds is slower or nil in lower soil horizons, due to diminished microbiological activity, herbicides reaching the subsoil are liable to remain phytotoxic for a longer period of time than in surface horizons. If the herbicides then enter ground water which is pumped to fields, susceptible plants could be affected. 2,4-D has been found to contaminate ground water at phytotoxic levels in the U.S. (Walker, 1961). Aquifers underlying irrigated farmlands which supply water to these lands and which are recharged from downward percolating water from fields and canal seepage, as in the Indus Valley, should be checked periodically for contamination by water-soluble herbicides.

The irrigation project planner cannot be expected to foresee all the management problems associated with agricultural chemical use. However, certain techniques of water application and regulation which are capable of alleviating contamination of water by herbicides and other pesticides might well be considered at the planning stage. For example, subsurface irrigation by means of perforated plastic pipes has been considered as a means of reducing salt accumulation (Law and Witherow, 1971) and might similarly reduce leaching of herbicides and pesticides. Again the contamination of ground water by these chemicals, through seepage from canals and irrigation channels, might possibly be reduced by lining channels with concrete. This would have the further advantages of impeding the spread of rooted waterweeds and lessening the need for the application of herbicides.

Finally, of course, the relation between water quality and water-borne diseases is as relevant to irrigation projects as it is to reservoir projects. The considerations discussed in section 4(c) of this chapter are equally pertinent.

(e) Efficiency of Water Use

As already mentioned, the application of excessive amounts of water to irrigated crops is a common failing, usually due to technical and managerial inexperience. As it can easily jeopardize the production goals of a project, the implications for the development planner are, first, that technical and management procedures to ensure the most efficient possible use of the water resource must be worked out and included in the plans and, secondly, that any necessary assistance or supervision to ensure that efficient operation is maintained, must be provided.

Apart from the economic reasons for efficient water use, unfavourable environmental conditions may result from applying water in excess of crop needs. The groundwater table may rise, producing salinization and soil waterlogging problems. Besides the toxic effect on crops of high concentrations of salts, saline water contains less dissolved oxygen and this may have adverse effects on aeration in the root zone, and on desirable organisms in the soil or in canals receiving saline runoff.

Calculations of water requirements for different crops at different stages of growth have been made, and general formulae have been worked out for potential evapotranspiration rates, which are approximately the same over the long term, regardless of vegetative cover; the main variables affecting evapotranspiration potential are temperature and rainfall. Water requirements for different crops do vary, however, according to stage of growth and resistance of plants to drought, and are subject to the losses in conveyance and application illustrated in Figure 31.

Field experience and climatic data may sometimes be lacking or inade-

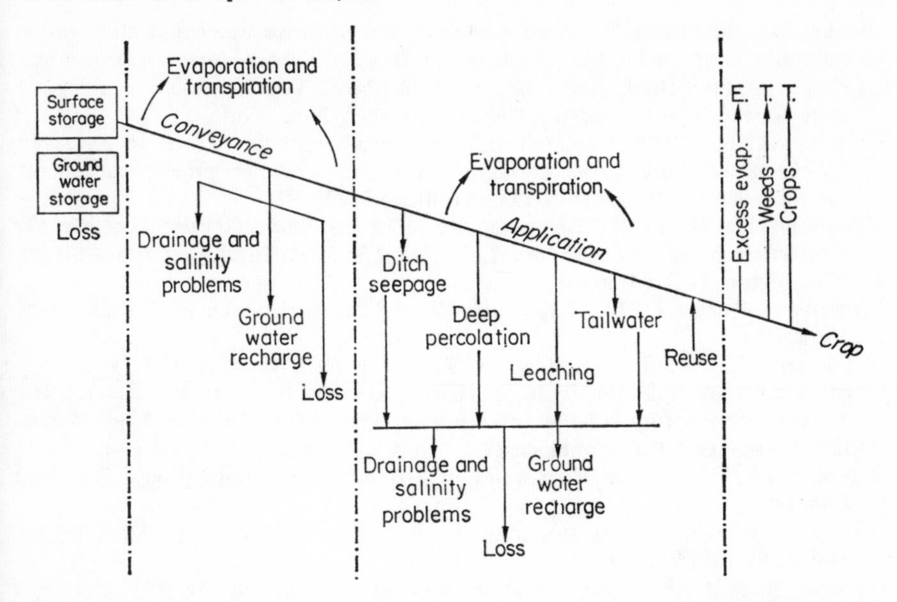

Figure 31. Processes and types of water losses occurring in conveyance, in application of irrigated water, and during crop production. (From Robert M. Hagan, 'Water management'. Figure 3, Publication 93. Copyright © 1970 by the American Association for the Advancement of Science).

quate for calculating water requirements and project capacity, especially in arid zones which have previously been considered of little agricultural value. At the preinvestment stage of irrigation projects, therefore, field trials are needed to determine water use rates for various crops in representative soils, in order to establish guidelines for water scheduling and arrive at realistic estimates of the total irrigable area.

(6) REFERENCES

Achi, Kamel (1972). Salinization and water problems in the Algerian Northeast Sahara. In *The Careless Technology: Ecology and International Development.* M. T. Farvar and John P. Milton, eds., Doubleday and Co., Natural History Press, New York.

Allen, Robert (1972). The Anchicaya hydroelectric project in Columbia: design and sedimentation problems. In *The Careless Technology: Ecology and International Development.* M. T. Farvar and John P. Milton, eds., Doubleday and Co., Natural History Press, New York.

Boffey, Philip (1971). Herbicides in Vietnam: AAAS study finds widespread devastation. *Science*, **171**, No. 3966.

Boyko, H. (1967). Salt-water agriculture. *Scientific American*, **216 (3)**, 89–94.

Boyko, H., and Boyko, E. (1968). Plant growing with sea-water and other saline waters in Israel and other countries. In *Saline Irrigation for Agriculture and Forestry.* Hugo Boyko (ed.), Dr. W. Junk, N.V., The Hague.

Brown, Lester (1970). *Seeds of Change.* Praeger. New York.

Bunting, A. H. (1970). Research and food production in Africa. In *Research for the World Food Crisis.* Daniel G. Aldrich, ed., American Association for the Adv. of Sci., Washington, D.C. Publication No. 92.

Challinor, David (1969). *Effects of the Mekong Basin Development on the Vegetation of the Forests and Lakes of Thailand.* Smithsonian Institution, Washington, D.C. Mimeo.

Chambers, Robert (1970). *The Volta Resettlement Experience.* Praeger, New York.

Chapman, V. J., Mill, C. F., Carr, J. T., and Brown, J. M. A. (1971). Biology of weed growth in the hydro-electrical lakes of the Waikato River, New Zealand. *Proceedings of the International Symposium on Man-Made Lakes,* Knoxville, Tenn. (to be published)

Clark, Colin (1970). *The Economics of Irrigation.* Pergamon Press. (2nd ed.) London.

Coleman, F. (1967). Irrigation and mechanization in the clay plains of the Sudan. *World Crops,* **17**, 3.

Cooper, Charles F. (1969). *The Ecosystem Concept in Natural Resource Management.* Academic Press, New York.

Courtney, K. Diane, *et al.* (1970). Teratogenic evaluation of 2, 4, 5-T. *Science,* **168**, No. 3933.

Farvar, M. T., and Milton, John P., eds. (1972). *The Careless Technology: Ecology and International Development.* Doubleday and Co., Natural History Press. New York.

Geertz, Clifford (1968). *Agricultural Involution.* University of California Press. Berkeley.

George, Carl (1972). The role of the Aswan High Dam in changing the fisheries of the southeastern Mediterranean. In *The Careless Technology: Ecology and International Development.* M. T. Farvar and John P. Milton, eds., Doubleday and Co., Natural History Press, New York.

Greany, W. H. (1952). Schistosomiasis in the Gezira irrigated area of the Anglo-Egyptian Sudan. *Annals of Tropical Medicine and Parasitology,* **46**, 250–67.

Greenman, D. W., Swarzenki, W. V., and Bennett, G. D. (1967). *Groundwater Hydrology of the Punjab, West Pakistan, with Emphasis on Problems Caused by Canal Irrigation.* Geolog. Survey water supply paper 1608-H.V.S. Government Printing Office, Washington, D.C.

Hagan, Robert M. (1970). Water management. In *Research for the World Food Crisis.* Daniel G. Aldrich, ed., Amer. Assoc. for Adv. of Sci., Publ. 92, Washington D.C.

Houston, Clyde E., and Allison, Stephen V. (1968). *Successful Irrigation.* FAO, Rome.

Hay, John (1972). Salt and salinity on the Upper Rio Grande. In *The Careless Technology: Ecology and International Development.* M. T. Farvar and John P. Milton, eds., Doubleday and Co., Natural History Press, New York.

Holm, L. G., Weldon, L. W., and Blackburn, R. D., 1969. Aquatic weeds. *Science,* **166**, Nov. 1969.

Hunter, John M. and Hughes, Charles C. (1972). The role of technological development in promoting disease in Africa. In *The Careless Technology: Ecology and International Development*. M. T. Farvar and John P. Milton, eds., Doubleday and Co., Natural History Press, New York.

Jackson, P. B. N. (1966). The establishment of fisheries in man-made lakes in the tropics. In *Man-Made Lakes,* R. H. Lowe-McConnell, ed., Symposia of the Institute of Biology No. **15**. Institute of Biology and Academic Press, London.

Jacobsen, Thorkild and Adams, Robert M. (1958). Salt and silt in ancient Mesopotamian agriculture. *Science*, **128**, No. 3334.

Kassas, Mohammed Abdul-Fattah, *et al.* (1972). Impact of river control schemes on the shoreline of the Nile delta. In *The Careless Technology: Ecology and International Development*. M. T. Farvar and John P. Milton, eds., Doubleday and Co., Natural History Press, New York.

Kharchenko (1970). *Symposium on World Water Balance*, Vol. II, pp. 454–5, UNESCO, Paris.

King, John A., Jr. (1967). *Economic Development Projects and their Appraisal: Cases and Principles from the Experience of the World Bank.* Johns Hopkins Press, Baltimore, Md.

Kneese, Allen V., and Smith, Stephen C. (1966). *Water Research.* Johns Hopkins Press, Baltimore, Md.

Kok, L. T. (1972). Toxicity to tropical fish in rice paddies by insecticides used for Asiatic rice borer control. In *The Careless Technology: Ecology and International Development*. M. T. Farvar and John P. Milton, eds., Doubleday and Co., Natural History Press, New York.

Lagler, Karl, ed. (1969). *Man-made Lakes: Planning and Development.* FAO and UNDP, Rome, Italy, 1969.

Law, James P., and Witherow, Jack L. (1971). Irrigation residues. *J. Soil and Water Conservation*, **26**, No. 2.

Leentvaar, P. (1971). Lake Brokopondo. *Proceedings of the International Symposium on Man-Made Lakes, Knoxville, Tenn.* (to be published)

Little, E. C. S. (1966). The invasion of man-made lakes by plants. In *Man-Made Lakes*, R. H. Lowe-McConnell, ed., Symposia of the Institute of Biology No. 15. Institute of Biology and Academic Press, London.

McGinnies, William G., Goldman, Bram J., and Paylore, Patricia (1968). *Deserts of the World. An Appraisal of Research into their Physical and Biological Environments.* University of Arizona Press, Tucson.

Michel, Aloys A. (1972). The impact of modern irrigation technology in the Indus and the Helmand Basins of Southeast Asia. In *The Careless Technology: Ecology and International Development.* M. T. Farvar and John P. Milton, eds., Doubleday and Co., Natural History Press, New York.

Milton, John P. (1969). *Pollution, Public Health and Nutrition Effects of Mekong Basin Hydro-Development.* Smithsonian Institution, Washington, D.C. Mimeo.

Milton, John P. (1971). *Ecology of a Tropical River Basin: A Mekong Case History*, The Conservation Foundation, Washington, D.C.

Muzik, Thomas J. (1970). *Weed Biology and Control.* McGraw-Hill Book Company, New York.

National Academy of Sciences, Committee on Water (1966). *Alternatives in Water Management*. Washington, D.C.

Organization of American States (1969). *Cuenca del Rio de la Plata: Estudio para su Planificacion y Desarrollo.* Inventario de datos hidrologicos y climatologicos. Washington, D.C.

Penman, H. L. (1970). The water cycle. *Scientific American*, **223**, No. 3.

Pirie, N. W. (1970). Weeds are not all bad. *Ceres, FAO Review*, **3**, No. 4.

Raheja. P. C. (1968). Saline soil problems with particular reference to irrigation with saline water in India. In *Saline Irrigation for Agriculture and Forestry.* Hugo Boyko, ed., Dr. W. Junk, N.V. The Hague.

Rahman, M. (1967). Problems of irrigation drainage salinization and water logging in the Sind region of West Pakistan. *Geography Rundschau*, **19**, 7.

Rothé, J. P. (1971). Earthquakes and filling of lake reservoirs. *Proceedings of the International Symposium on Man-Made Lakes*, Knoxville, Tenn. (to be published).

Schuphan, W. (1972). Nitrate problems and nitrite hazards as influenced by ecological conditions and by fertilization of plants. In *The Careless Technology: Ecology and International Development.* M. T. Farvar and John P. Milton, eds., Doubleday and Co., Natural History Press, New York.

Scudder, Thayer (1972). Ecological bottlenecks and the development of the Kariba Lake basin. In *The Careless Technology: Ecology and International Development.* M. T. Farvar and John P. Milton, eds., Doubleday and Co., Natural History Press, New York.

Scudder, Thayer (1966). Man-made lakes and population resettlement in Africa. In *Man-Made Lakes*, R. H. Lowe-McConnell, ed., Symposia of the Institute of Biology No. 15. Institute of Biology and Academic Press, London.

Sen, B. R. (1967). Use of water resources. *Development Digest*, **5**, No. 3.

Shiff, C. J. (1972). The impact of agricultural development on aquatic systems and its effect on the epidemiology of schistosomes in Rhodesia. In *The Careless Technology: Ecology and International Development.* M. T. Farvar and John P. Milton, eds., Doubleday and Co., Natural History Press, New York.

Shiff, C. J. (1972). The effects of molluscicides on the microflora and microfauna of aquatic systems. In *The Careless Technology: Ecology and International Development.* M. T. Farvar and John P. Milton, eds., Doubleday and Co., Natural History Press, New York.

Szestay, K. (1971). Hydrology and man-made lakes. *Proceedings of the International Symposium on Man-Made Lakes, Knoxville, Tenn.* (to be published).

Talbot, Lee M. (1970). *Effect of Mekong Development on Biotic Factors, Particularly Wildlife, Parks and Reserves.* Smithsonian Institution, Washington, D.C. Mimeo.

Taylor, Carl E., and Hall, Marie-Françoise (1967). Health, population and economic development. *Science*, **157**.

Thomas, Harold E. (1956). Changes in quantities and qualities of ground and surface waters. In *Man's Role in Changing the Face of the Earth.* William L. Thomas, Jr., ed., University of Chicago.

Timmer, C. E., and Weldon, L. W. (1967). *Hyacinth Control J.*, **6**, 34 (1967); W. T. Penfound and T. T. Earle, *Ecological Mongr.*, **18**, 447 (1948); R. R. Das, *Proc. Indian Sci. Congr.*, **6**, 445 (1969).

United Nations (1970). *Integrated River Basin Development.* New York.

United States Department of Agriculture (1965). *Waterweed Control on Farms and Ranches. (Farmer's Bulletin No. 2181).* United States Government Printing Office. 21 pp.

United States President's Science Advisory Committee (1971). Panel on Herbicides. *Report on 2, 4, 5-T.* United States Government Printing Office, Washington, D.C. 68 pp.

van der Schalie, Henry (1972). World Health Organization Project Egypt 10: a case history of a schistosomiasis control project. In *The Careless Technology: Ecology and International Development.* M. T. Farvar and John P. Milton, eds., Doubleday and Co., Natural History Press, New York.

van der Schalie, Henry (1968). Control in Egypt and Sudan. In *The Unforeseen Ecological Boomerang.* M. T. Farvar and John P. Milton, eds., Natural History Magazine, New York.

Waddy, B. B. (1966). Medical problems arising from the making of lakes in the tropics. In *Man-Made Lakes,* R. H. Lowe-McConnell, ed., Symposia of the Institute of Biology No. 15. Institute of Biology and Academic Press, London.

Walker, R. T. (1961). Groundwater contamination in the Rocky Mountain Arsenal Area, Denver, Colorado. *Geological Society of America, Bulletin No. 72,* pp. 489–94.

White, E. (1969). Man-made lakes in tropical Africa and their biological potential. *Biological Conservation.*

White, Gilbert (1972). Organizing scientific investigations to deal with environmental impacts. In *The Careless Technology: Ecology and International Development.* M. T. Farvar and John P. Milton, eds., Doubleday and Co., Natural History Press, New York.

Williams, C. N., and Joseph, K. T. (1970). *Climate, Soil and Crop Production in the Humid Tropics.* Oxford U. Press, Singapore.

Index